# THE HIGH FRONTIER

## Human Colonies In Space

### 3rd Edition

### Gerard K. O'Neill

with contributions by
David P. Gump, Margo R. Deckard, M.S.,
Peter E. Glaser, Ph.D., John S. Lewis, Ph.D.,
Rick N. Tumlinson, and George Friedman, Ph.D.

Presented by
the Space Studies Institute,
the Space Frontier Foundation,
and Apogee Books

*To the O'Neill grandchildren*
*Niko*
*Luke*
*and Ian*

We acknowledge the financial support of the Government of Canada through
the Book Publishing Industry Development Program for our publishing activities.

Published by Apogee Books in conjunction with
the Space Studies Institute and the Space Frontier Foundation

Apogee Books is an imprint of Collector's Guide Publishing Inc.,
Box 62034, Burlington, Ontario, Canada, L7R 4K2

Printed and bound in Canada

The High Frontier, 3rd edition

by Gerard K. O'Neill

ISBN 1-896522-67-X 3rd edition © 2000 Space Studies Institute

(1st Edition ISBN 0-688-03133-1 © 1976 Gerard K. O'Neill)
(published by William Morrow and Company Inc., 1977)

(2nd Edition ISBN 0-9622379-0-6 © 1976, 1977, 1982 Gerard K. O'Neill, 1989 SSI)
(published by Space Studies Institute Press, 1989)

Cover Design and Artwork:  Peter Thorpe

Cover Illustrations:  David Lauterbach

Picture Credits:  NASA, SSI, page 82 Mark Martel, page 100 Pat Rawlings

# *Table of Contents*

# Preface

It was a quarter century ago that Gerard O'Neill first published *The High Frontier*. Freeman Dyson points out in his introduction to this new edition that Gerry O'Neill's vision for humanity's migration into space has not *yet* come to pass. It's important to realize though that the inspiration we may gain from the imaginations of visionaries is more important, in terms of driving humanity forward, than is the ability of the visionaries to accurately predict the timeline upon which their vision will unfold.

In 1976, a central message of *The High Frontier* was that the technology already existed to build the space colonies, mass drivers and solar power satellites described in the book. Today, at the beginning of the new millennium, that message is no less true. In several respects technological advances of the last 25 years have further enabled the original vision. However, the human species has collectively not chosen to *develop* those technologies in a way and to an extent such that we are appreciably closer to the High Frontier today. We have also collectively shrunk away from the realization that fossil energy is a dwindling and environmentally dirty resource. World population *has* continued to grow at an alarming rate. Twenty-five years ago we didn't realize that the mass extinctions which occurred 65 million years ago were the result of an impact by an object from space. We now realize that such objects are in fact relatively common, and that impacts with objects much larger than that which eliminated the dinosaurs are not only possible, but also remarkably likely in the lifetime of our planet.

Notwithstanding intriguingly suggestive microscopic and chemical data from a meteorite collected several years ago in the Antarctic, we still possess no convincing evidence that life exists anywhere else in the universe. This means that, as far as we know, if we fail to preserve life on Earth through its expansion off planet, we're placing in jeopardy the very existence of life, anywhere, forever. One conclusion from all of the above is that not only does the technology still exist to implement "the vision" of *The High Frontier*, the same compelling reasons still exist to do so.

Why produce a new edition of *The High Frontier*? An 18-year-old who read the first edition in 1976 would be 42 years old today. I was such an 18-year-old. A whole new generation of creative, forward thinking 18-year-olds exist today who were not yet born when the first edition was published. Meanwhile, many people who were inspired by the first edition in 1976 are still working towards the vision, in a large variety of small ways. These include members of the Space Studies Institute (founded by Gerard and Tasha O'Neill in 1977); the Space Frontier Foundation; many other space advocacy groups; and a wide variety of scientific, engineering and other occupations that support in many ways, large and small, humanity's eventual migration off planet. These people will find refreshing the new sections of this latest edition of *The High Frontier*.

New sections in this edition include an introduction by Freeman Dyson, who points out that launch costs for materials from the Earth's surface are probably the single factor most limiting the realization of the vision of *The High Frontier*. At the same time, he identifies a technology in its infancy that may provide a solution to this problem — light-powered spacecraft.

New chapters by Peter Glaser (Space Solar Power for the 21$^{st}$ Century) and Margo Deckard (A Technology For A Better Future: Space Solar Power) update and advance the case for solar power satellites, with additional research-based information unknown in 1976. John Lewis has contributed a chapter (Asteroid Resources, Exploitation, and Property and Mineral Rights, *or*, Keep Your Laws off My Asteroid.) This speaks to the fact that near-Earth object (NEO)-based resources have gained favor over the Moon-based resources described in the first edition as raw materials for large, space-based construction and manufacturing projects. Also included as a new chapter is Rick Tumlinson's A Conspiracy Of Dreamers, following closely his presentation in the spring of 1999 to the Space Studies Institute-sponsored conference on space manufacturing, held at Princeton University. David Gump's Space Robotics chapter details the role that advanced robots will play in exploring and developing space and its resources. And a new chapter, At Millennium's Eve, by George Friedman, Executive Vice President of the Space Studies Institute, provides a soberingly realistic look back on what we have learned in the 25 years since the first publication of *The High Frontier*. George's chapter is at the same time compellingly optimistic concerning how what we have learned continues to build the case for *The High Frontier*, and provides new ways of thinking about how we may get there.

In closing, I am hopeful that this new edition will provide first exposure for a new generation, while at the same time satisfying the desire for evolution and updating of the ideas expressed in the first edition. I have no doubt that some day the vision of Gerard O'Neill will be realized — the only question lies in which generation of our species will finally see it happen.

Roger O'Neill
Chairman, Space Studies Institute
January 20, 2000

# Introduction

We who believe in the High Frontier as a real part of the human future have to face the fact that it didn't happen the way Gerard O'Neill described it in this book. If we do not face this fact, we have no chance of persuading the skeptics. O'Neill described Satellite Solar Power Stations transmitting electrical power to the Earth and generating enough revenue to attract the massive investment of private funds. He described mining operations on the Moon, and manufacturing operations at the place called L5 in space where permanent structures easily stay put. He described families traveling up to live at L5 in spacious and comfortable habitats. None of these things have happened and none are likely to happen in the next twenty years.

O'Neill was careful not to predict a definite timetable for the things he described, but he certainly expected that the first large scale human settlement in space would be occupied between 1995 and 2010. We now know that this will not happen. The International Space Station that is supposed to be built during this period is in no way a fulfilment of O'Neill's dream. It will not be commercially profitable; it will not be producing anything of value commensurate with its cost; and it will not be a spacious and comfortable habitat. It will be in low Earth orbit instead of at L5, so that it will have no access to lunar materials. The idea of a young family moving into the International Space Station to enjoy the amenities of the High Frontier is manifestly absurd. The International Space Station is not a step forward on the road to the High Frontier. It's a big step backward, a setback that will take decades to overcome. It's no wonder that the general public, seeing the obvious irrelevance of the International Space Station to human needs, concludes that O'Neill's dream is equally irrelevant. It is up to us, the believers in the dream, to explain what went wrong and why the dream still makes sense.

What went wrong was O'Neill's expectation that NASA would provide the infrastructure to help him do what was needed. In the beginning, when he launched his plans for space colonization, he had the choice of two roads to follow — he could work with NASA, or he could work independently of NASA. He chose to work with NASA. He accepted NASA funds to support some of his projects, and he accepted the NASA Space Shuttle and its derivatives as launchers for his future missions. Like other large bureaucratic organizations, NASA has many parts that are only loosely connected with one another. There is the real NASA, that builds and operates the Shuttle and the Space Station, and there is the paper NASA, that studies advanced concepts and long range ventures. The real NASA is intensely conservative, dedicated to preserving the existing programs, and especially dedicated to preserving the Shuttle and the Space Station. The paper NASA is adventurous on paper but lacks the resources to be adventurous in reality. The paper NASA supports advanced concepts as long as they're far in the future, but cannot develop alternative infrastructures that might compete seriously with the Shuttle. When O'Neill appeared on the scene, his plans were far-out enough to fit well with the style of the paper NASA. The paper NASA was glad to fund studies of his ideas and to give him public visibility. But the real NASA never took his ideas seriously and never gave him the help that he needed.

O'Neill's dream makes sense if, and only if, the cost of launching stuff from the ground into space can be drastically reduced. Present costs, using either the Shuttle or other available launchers, are roughly ten thousand dollars a pound. Estimated costs for launchers now being developed by aerospace companies are roughly a thousand dollars a pound. These costs are still to be demonstrated, and are probably as low as can be expected as long as we continue to use chemical rocket launchers. The next big jump after that will require a "public highway" launch system, with a stationary machine on the ground supplying the energy, so that the payloads to be launched don't need to lift the weight of their own fuel. A public highway system will only be cost effective if it has a large volume of traffic to keep it busy. Given a large enough volume of traffic, the launch costs of a public highway system could conceivably come down below a hundred dollars a pound. The public is well aware that with present day launch costs human activity in space must remain a spectator sport, with a small team of elite performers paid for by the crowds of people who stay on the ground and watch the show on television. O'Neill's dream was to open the sky to the crowds, to allow ordinary people to homestead the asteroids. He dreamed that ordinary people would be participants rather than spectators.

Ordinary people can easily calculate that their chance of being able to afford a trip into space at present day costs is equal to their chance of winning a big jackpot in a lottery. A minimal payload for a family emigrating to L5 would be about five tons. In my opinion, ordinary people will take O'Neill's dream seriously as soon as the cost of emigration for a family falls below a million dollars. At that price, only a small fraction of the population will have a chance of going, but the idea will no longer seem absurd. At present day prices, the cost of emigration would be at least a hundred times too high, even if there were a place to go to where people could survive. With the next generation of rocket launchers, the cost will still be ten times too high. But with a public highway system the cost might be reasonable.

In April of 1997, the thirteenth Space Manufacturing Conference was held at Princeton, organized by the Space Studies Institute which O'Neill founded in 1977. Many of O'Neill's old friends and supporters were there. We saw an historic event that might be our first step into a new world. Professor Leik Myrabo reported on his work and showed a film of his first test flight. He is a professor at Rensselaer Polytechnic Institute and is developing a laser propelled spacecraft which he calls Lightcraft. In the crucial early phases, his work was supported by the Space Studies Institute. More recently, he has received substantial help from the United States Air Force.

The concept behind the Lightcraft is using a laser beam as a public highway from the ground into space. The laser stays on the ground and provides the power. The Lightcraft flies up the beam, using the laser to power a rocket motor using ordinary water or air as its propellant. Until 1997, nobody had succeeded in building a laser powered motor with enough thrust to lift itself off the ground. Myrabo's film showed the first flight of a little model vehicle that he calls the Lightcraft Technology Demonstrator. It rose three feet into the air, not as high as the first flight of the Wright brothers in December 1903. More recent Lightcraft flights flew higher, but the first flight was the decisive step. The first flight demonstrated that laser propulsion is possible, just as the 1903 flight demonstrated that human heavier-than-air flight was possible. The sound track accompanying the film made a noise like a machine gun, as the laser fired high intensity pulses at a rate of ten per second. The vehicle has a little fish shaped body with a blunt nose and a shiny reflecting dish around its waist. It weighs two ounces and has a diameter of six inches. Each laser pulse is focussed by the dish and heats the air at the focus to a high temperature, causing a shock wave that propels the model upward. The power in the beam is ten kilowatts. The laser was borrowed from the US Air Force at white Sands, New Mexico.

It took fifty years to go from the Wright brothers' *Flyer One* of 1903 to the modern air transport system with huge numbers of commercial aircraft flying routinely all over the world. Perhaps it will take another fifty years to go from Myrabo's Lightcraft Technology Demonstrator to a laser propelled public highway system with huge numbers of spacecraft traveling routinely to destinations in geostationary orbit, on the Moon, at L5 and beyond. At each destination there must be a massive infrastructure comparable with a major airport and including the associated industries and hotels. All this could be done in fifty years, if and only if the launch costs of the public highway into space come down into the range of present day costs of intercontinental air travel.

Laser propulsion is not the only new technology that might lower launch costs drastically. If laser propulsion fails to achieve its promise, other methods of building public highways into space are possible. Laser propulsion works with modest acceleration and aims to carry human passengers. The technical problems of the launch system are less severe if it is used only for freight and not for passengers. Then systems with very high acceleration are possible — modern versions of Jules Verne's mythical cannon that launched the president and secretary of the Baltimore Gun Club into orbit around the Moon. Two other promising systems are the Ram Accelerator, a gentler and more efficient high velocity gun, and the Slingatron, a gigantic slingshot machine accelerating a payload around a circular track. Both these systems, like Myrabo's Lightcraft, have been demonstrated with small scale models that performed well. Derek Tidman demonstrated his Slingatron at the twelfth Space Manufacturing Conference organized by the Space Studies Institute in 1995. It would not be surprising if all three systems could be developed in the next fifty years to provide public highways serving different users with different needs and different kinds of cargo.

I hope that the majority of readers of this book will be skeptics rather than believers in the O'Neill dream. For believers, the book may tend to raise unrealistic expectations. Believers should be warned that the infrastructure described by O'Neill in 1976, with launches provided by the Shuttle and other Shuttle-related vehicles, no longer makes sense. Believers must accept the fact that O'Neill's reliance on NASA to bring his dream to reality was a grand illusion. Believers must resign themselves to a long delay, probably as long as fifty years, before the dream will become practical. But the real value of this book is the message it will bring to skeptics. Skeptics need to be convinced that the O'Neill dream can still make sense as a realistic future for our species, whenever in the fullness of time we shall deploy a new technology that reduces launch costs by a factor of a hundred. Skeptics will find in this book a variety of ingenious ideas and imaginative scenarios, all of them making sense if the crucial cost-barrier can be overcome. Skeptics may read the book as a work of science fiction, suspending their disbelief long enough to enjoy the drama that O'Neill describes. Finally, you don't have to be a believer to read this book as a monument to a great man whose dream, like the dream of Martin Luther King, may have the power to move us into a new world long after the dreamer has died.

Freeman J. Dyson
President, Space Studies Institute

# *As You Read This Book*

This third edition of *The High Frontier* is presented in two parts.

Part One includes essentially the complete text of the second edition as published in 1989, which itself was only slightly updated from the original edition published in 1977. The concepts in *The High Frontier*, and the way in which they're presented, are such that no major revisions of O'Neill's vision have been necessary. Certainly new knowledge requires us to modify the details of implementation, but the vision itself is as valid and as important today as it was in the 70's, when Gerry O'Neill did his original research — with each passing year it becomes ever more important to the collective welfare of the human race.

Part Two provides six new chapters, reflecting the work done and the knowledge gained in the last 25 years. These chapters have been written by people who are not only recognized as leaders in their respective fields, but who have also been involved in the active pursuit of the High Frontier for many years. What these new chapters clearly show is that today we still have the technology we need to reach and hold the High Frontier, and more reason than ever to do so. All that's missing is the commitment. And the message is clear — the time is now!

# Part One
# The Vision

# Chapter 1
# *A Letter From Space*

In 1974 a new concept to improve the human prospect entered the arena of open discussion. Its thrust is to open for our use new sources of energy and materials while preserving our environment. First it was known as "space colonization," but now, as it is discussed with increasing seriousness in the circles of government, business, the universities and the press, we tend to use for it less dramatic names: "space manufacturing," or "high-orbital manufacturing."

The concept of human habitation in space is, of course, a very old one. In some form it can be traced back to the early days of science, and even earlier to mysticism. It has been a theme for fiction over several decades, and at least one fictional discussion of an inhabited artificial satellite, by Edward Everett Hale, was written during the latter half the nineteenth century. The Russian schoolmaster and physicist Konstantin Tsiolkowsky foresaw certain elements of the space community concept with remarkable clarity. In his novel, *Beyond the Planet Earth*, written about 1900 and published some twenty years later, Tsiolkowsky set his space travelers to work, on their very first voyage, constructing greenhouses in space beyond the Earth's shadow, and there raising crops to support a population of émigrés from the Earth. His astronauts visited the Moon, but only as an excursion in passing — their most important destination was the asteroids, a vast resource of materials.[1-6]

Still other authors, most of them writing later in the twentieth century, played with the idea of habitats in space. Lasswitz in 1897, Bernal, Oberth, Von Pirquet, and Noordung in the 1920's, continued the theme,[7-14] as did Wernher von Braun, Dandridge Cole, and Krafft Ehricke in the 1950's and 1960's.[15-26] Although many of these ideas are echoed in this book, it would have been difficult, before the year 1969, to make of them a coherent picture without serious technical gaps.

Our goal is to find ways in which all of humanity can share in the benefits that have come from the rapid expansion of human knowledge, and yet prevent the material aspects of that expansion from fouling the worldwide nest in which we live. Necessarily, many of the concerns of this book are materialistic, but more than material survival is at stake. The most soaring achievements of mankind in the arts, music and literature could never have occurred without a certain amount of leisure and wealth. We should not be ashamed to search for ways in which all of humankind can enjoy that wealth.

The humanization of space is both an old dream and a present day reality. As living space, agricultural area and manufacturing facilities, our future is in space.

A firm schedule for the development of resources in space would depend on decisions not yet made, but it appears that construction of a high-orbital facility could begin within seven to ten years using launch vehicles no more advanced than those of today [1976], and that it could be completed within fifteen to twenty five years.

Governmental interest in high-orbital manufacturing stems in part from calculations on its economics. These suggest that a community in space could supply large amounts of energy to the Earth, and that a private, perhaps multinational investment in the first space habitat could be returned several times over in profits.

Much of the public interest relates to the human prospect that thousands of people now alive may choose within the next decades to live and work on a new frontier in space. If this concept is realized as soon as is technically possible, something like the following "letter from space" might be written within the next twenty years.

*Dear Brian and Nancy:*

*I can understand why you want to hear from someone who's working and living in space before deciding whether to make the commitment yourselves. According to your letter you've reached the "finals" in the selection process now. The next step will be the admission interview. After that, if you get an offer, you'll have to decide whether to go for the six months' training. Though I was never in the Peace Corps, I understand that the selection methods are similar to theirs. Most people in your training group will pass the tests.*

*Then there's the big step of the first spaceflight, the three week stay in orbit. By now the flight itself is quite routine. You'll find that the single stage shuttle interior is much like that of one of the smaller commercial jets. There'll be one hundred and fifty of you traveling together. The G forces will be higher than in commercial aviation, but still nothing to worry you. The trip into orbit will only take about twenty minutes, and then you'll experience something really new: zero gravity. You may feel queasy at first — as if you were on a ship at sea. The three week trial period is to sort out cases of severe space sickness, and to find out whether you are among those who can adapt to commuting each day between normal gravity and zero. That's important because our homes are in gravity obtained by rotation, and many of us work in the construction industry, with no gravity at all. Those who can adapt to rapid change qualify for higher paying jobs. The trial period also gives people the chance to decide that "this is not for me."*

*After three weeks you'll be ready to transfer to one of the "liners" on its next trip in. Jenny and I enjoyed that voyage. You'll be on the Goddard or the Tsiolkowsky, and each takes a week for the outbound passage. About half of the passengers will be newcomers like yourself, and half will be returnees coming back from vacations on Earth. The ship rotates, so there will be gravity, normal in the public rooms and less than that in the sleeping cabins. In the six months' training period you'll have had cram courses in foreign languages, so try talking with some of your fellow passengers from other countries. We like visiting nearby communities for dinners out fairly often, and enjoy talking with people we meet there even though our foreign language ability is mainly of the "restaurant" variety.*

*In space near the communities, the biggest things you'll see will be solar satellite power stations being assembled to supply energy for Earth. Those power stations are about ten times as big as the habitats themselves. You won't see much detail from the outside of the habitats because they're shielded against cosmic rays, solar flares, and meteoroids by a thick layer of material, mainly slag from the processing industries.*

*All the habitats are variations of basic sphere, cylinder or ring shapes. We live in Bernal Alpha, a sphere about five hundred meters in diameter, with a circumference inside, at its "equator" of nearly a mile. We have track races and bicycle races that use the ring pathway. That path wanders all the way round, generally following the equator, and near it is our little river. Bernal Alpha rotates once every thirty two seconds, so there is Earth gravity at the equator. The land forms a big curving valley, rising from the equator to 45 degree "lines of latitude" on each side. The land area is mainly in the form of low-rise, terraced apartments, shopping walkways, and small parks. Many services, light industries, and shops are located underground or in a central low gravity sphere, or are steeply terraced, because we like to preserve most of our land area for grass and parks. Our sunshine comes in at an angle near 45 degrees, rather like mid-morning or mid-afternoon on Earth. The day length and therefore the climate are set by our choice of when to admit sunlight. We keep Canaveral time, but two other communities near us are on different time zones. All the communities serve the same industries, so the production operations run twenty four hours a day, three shifts, but with no one having to work the night shift.*

*Alpha has a Hawaiian climate, so we lead an indoor-outdoor life all year. Our apartment is about the same size as our old house on Earth, and it has a garden. Alpha was one of the first habitats to be built, so our trees have had time to grow to a good size.*

*You'll notice immediately the small scale of things, but for a town of 10,000 people we're in rather good shape for entertainment: four small cinemas, quite a few good small restaurants, and many amateur theatrical and musical groups. It takes only a few minutes to travel over to the neighboring communities, so we visit them often for movies, concerts, or just a change in climate. There are ballet productions on the big stage out in the low gravity recreational complex that serves all the residents of our region of space. Ballet in one-tenth gravity is beautiful to watch: dreamlike, and very graceful. You've seen it on TV, but the reality is even better. Of course, right here in Alpha we have our own low gravity swimming pools, and our club rooms for human powered flight. Quite often Jenny and I climb the path to the "North Pole" and pedal out along the zero gravity axis of the sphere for half an hour or so, especially after sunset, when we can see the soft lights from the pathways below.*

*You asked about our government, and that varies a great deal from one community to another. Legally, all communities are under the jurisdiction of the Energy Satellites Corporation (ENSAT) which was set up as a multinational profit making consortium under U.N. treaties. ENSAT keeps us on a fairly loose rein as long as productivity and profits remain high — I don't think they want another Boston Tea Party. There are almost as many different kinds of local government as there are national groups within the colonies. Ours happens to be a town meeting style. That wouldn't work in a town of as many as 10,000 people, except for the facts that all of us are much too busy to make a hobby of electioneering, and that the basics of habitat survival require a high level of competence on the part of the maintenance people. Our teenagers have to work a year in one of the life support maintenance crews — it's a little like military service on Earth — and if the regular government or maintenance people were to get balky, they'd be replaced by volunteers awfully fast.*

*Jenny and I laughed a bit about your comment on having to give talks to civic groups — I remember we went through the same things ourselves.*

*For information to use in your lectures, I'll mention a few basics. The initial stock of water for each habitat is obtained by combining hydrogen brought from Earth with eight times its weight in lunar oxygen. Here at L5, oxygen is a waste product from the industrial processes that turn out metals and glass. Our soil, of course, comes from the Moon and is fertile once we add water and nitrates. Because of our unlimited cheap energy, we don't have pollution here. Where energy costs almost nothing, and raw materials are relatively expensive, it pays to break down every waste product into its constituent elements.*

*So far there aren't enough communities to make long distance travel a problem, but when there are many of them, spaced over thousands of miles, we already know how the transport system will work. We can just accelerate an engineless vehicle to a high cruising speed by an electric motor at one community, and then after a trip of several thousand miles, we can slow it to a halt by an arresting cable at another community.*

*A long time ago someone calculated the maximum size for space habitats. They could be made in sizes at least as large as twelve miles in diameter, with a land area of several hundred square miles in each one. We're already talking about shifting the mining base from the Moon to the asteroids, where we'll have a complete range of elements including carbon, nitrogen, and hydrogen. In energy, it won't be any harder for us to get materials from the asteroids than from Earth, and it should be a lot cheaper because the transport system can take its time and won't ever need high thrust. Someone calculated how much "room for growth" there will be once we start to use the asteroidal material. The answer came out absurdly high: with the known unused materials out there, we could build space communities with a total land area 3,000 times that of Earth.*

*To go on with our situation, it's a comfortable life here. Fresh vegetables and fruit are in season all the time, because there are agricultural cylinders for each month of the year, each with its own day length. We grow avocados and papayas in our own garden and never need to use insecticide sprays. Of course we like being able to get a suntan without ever being bitten by a mosquito. To be free of those pests, it's worth it to go through the inspections before getting aboard the shuttle from Earth.*

*You asked whether we feel isolated. Some of us do get "island fever" to some degree, probably because we're really first generation immigrants — it never seems to bother the kids who were born here. When you sign your contract there are clauses that help quite a bit though. One is the provision for free telephone and videophone time to Earth. Another sets up free transportation to Earth and return on a space-available basis. Jenny and I took six months' leave after our first three years here. Our visit was luxurious, because our salaries are paid in part in Earth currency; we're both employed, Jenny as a turbine blade inspector and I in precision assembly. Our housing, food, clothing, and the rest are purchased in SHARES (Standard High-orbital Acquisition units Recorded Electronically) so our Earth salaries just accumulate in the bank. When we went back we had a lot of money to spend, and even on a luxury basis we couldn't go through it in six months.*

*We found something though, that may help to answer your basic question: by the time the vacation was nearly over, we were very ready to come back here. We missed our own place. Jenny is an enthusiastic gardener, and though other people were living in our apartment here and taking care of the greenery, she wanted to be at home to enjoy it herself. And I missed the friends I'd been working with. I can best describe the other thing that drew us back by saying that the space habitats are exciting places to be. They're growing and changing so fast that if you're away for six months you've missed a lot.*

*As to whether you'll really like it, of the people who came with us, more than half intend to stay after their five year contract is up. I understand that the settlement of Alaska has had about the same kind of "stay-ratio."*

*Now we're beginning to ask ourselves: will we want to retire to Earth or not? We don't have to face that for another twenty years, but we can see already that it won't be an easy decision. Some of us who are handy with tools have formed a club to design spacecraft for our own construction — rather like the homebuilt aircraft clubs on Earth. We're thinking of homesteading one of the smaller asteroids, and the numbers look reasonable. Especially if our daughter and son-in-law decide to come along, with the grandchildren, I think we're more likely to move further out than go back.*

*If you decide to come out, let us know what flight you'll be on and we'll meet you at the docks. We'd like you to come over to our place for supper, and we'd be glad to help you to get settled.*

*With our best wishes for good luck on the tests,*
*Cordially,*
*Edward and Jenny*

As we explore these possibilities we must remember that they are just that — not predictions or prophecies. The time scale may be longer than the fifteen to twenty five years I estimate to be an achievable minimum; or I may be too cautious, and events may dictate a still faster scale. The "when" is not science but a complicated, unpredictable interplay of current events, politics, individual personalities, technology and chance. As a guess, though, I consider it unlikely that the first community in space will be established in less than fifteen years, and also unlikely that it will be delayed for another fifteen years beyond that. Neither of these dates is very far off, and both are within the lifespan of most people now alive. In the matter of dates, it is to me rather thought provoking that Konstantin Tsiolkowsky, the great visionary space pioneer of nineteenth century Russia, was himself too conservative on the date of the first Earth-orbital flight: he guessed the year 2017.

Robert Goddard (1882-1945), much of whose life was spent in the more practical and therefore much more difficult task of reducing the theory of rocketry to working hardware, left us with a caution lest our vision be too narrow:

> "It is difficult to say what is impossible, for the dream of yesterday is the hope of today and the reality of tomorrow."

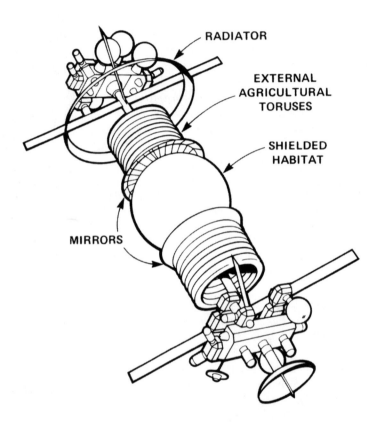

The "Bernal sphere" habitat design has a spherical living area bracketed by banded torus agricultural areas on either side. Non-rotating industries are at the ends of the axis.

# *Chapter 2*
# *The Human Prospect On Planet Earth*

We now have the technological ability to set up large human communities in space — communities in which manufacturing, farming, and all other human activities could be carried out. Substantial benefits, both immediate and long term, can accrue to us from a program of expansion into that new frontier.

The normal first reaction to such a statement is disbelief — isn't such a development beyond us? Not at all: the settlement of space by humans could be carried out without ever exceeding the limits of the technology of this decade. But even if it's possible, should we make the effort? I believe we should. The reasons go from an immediate and severely practical one: solving the energy crisis which we face here on Earth; to the slightly longer term problem of population size and Earth's capacity to support it; finally to a nonmaterial problem, compelling but not to be reckoned in dollars: the opportunity for increased human options and diversity of development.

Through many tens of thousands of years human beings were few in numbers, and insignificant in their power over the physical environment. Not only war but famine and plague decimated populations whenever they grew large. Centuries passed without great increase in the total human population. The quality of life, for most people in those preindustrial years, seems to have been low even in times of peace. Although there were, nearly everywhere, small privileged classes enjoying comparative wealth, most people lived out their lives in heavy labor, many as slaves.[1] Through all that time any observer on another planet would have found it very difficult to find telescopic evidence for the existence of the human race — our power over Earth was too slight to be noticeable.

Very suddenly, in a time of less than two hundred years, our human status as passengers on a giant planet, lost in its immensity and powerless before its forces, has changed dramatically. The beginnings of a science of medicine, and the rapid development of chemistry, have made fatal disease a rarity among children in the wealthy nations, and have even reduced its power in the poorer nations. With that one radical change we suddenly find ourselves growing in numbers so fast that Earth itself cannot long sustain our increase.

At the same time, our power to change the surface of Earth has increased: our activities can and now do alter the planet and its atmosphere. We achieve every year a greater degree of control over the natural environment, and we change it more in attempts to suit our liking. The result, though, does not always please us.

The industrial revolution has been the mechanism by which our physical power has increased, and by which, for the first time, a substantial fraction of the human population has reached a high living standard. Comfort, a reasonable life expectancy, freedom to travel, the easy availability of news and education — these have come to the most advanced countries as benefits of industrialization. But that process has brought evils as well. Though it began only two hundred years ago, less than a ten millionth of the time since Earth was formed, its side effects have already altered Earth in frightening ways. It has scarred, gutted, and dirtied our planet to a degree that many people find intolerable. Smoke and ash from factories in England cloud the air as far away as Norway, and pollutants from the industries of Japan can be detected in the snows of Alaska. Nearly every major city has its air pollution problem.

If those evils had occurred after the industrial revolution had penetrated to every nation on Earth, we could have tried to discuss, as a species, the actions necessary to counteract them. But we're not so fortunate — the evils of environmental damage are minor compared to others that have appeared: sharp limits on food, energy, and materials confront us at a time when most of the human race is still poor, and when much of it is on the edge of starvation. We can't solve that problem by retreat to a pastoral, machine-free society — there are too many of us to be supported by preindustrial agriculture. In the wealthier areas of the world, we depend on mechanized farming to produce great quantities of food with relatively little human effort; but in much of the world, only backbreaking labor through every daylight hour yields enough food for bare survival. About two thirds of the human population is in underdeveloped countries. In those nations only a fifth of the people are adequately fed, while another fifth are "only" undernourished — all the rest suffer from malnutrition in various forms.[2]

In those countries the need to increase the food supply is desperate. When the land cannot support its population, and starvation is general, disease strikes at the old and even harder at the young. Small children of a family contract the crippling diseases of malnutrition; parents must watch their children die, and be powerless to save them. In these areas some degree of industrialization is not a luxury but a desperate need. It is a great tragedy of the late twentieth century that the satisfaction of such a need is being denied or delayed in part because of the energy and materials limits of Earth.

As we view the process that has given most people in the industrialized world some freedom of movement and relief from heavy labor, we find that it's based on the increasing use of artificial energy sources. Within our own lifetimes we have seen rapid long distance travel become commonplace for a large fraction of the population — forty years ago it was impossible even for the very rich. A luxury passenger liner of the 1930's took several days to cross the Atlantic and its engines developed about twenty horsepower per passenger carried. Now, a crossing by jet aircraft takes only a few hours, but the plane needs several hundred horsepower per passenger. Until the energy crisis of 1973-74, energy usage in the United States was growing by 7 percent per year.[3] The mechanization of agriculture, the "green revolution," and the rapid development of nonfarming industry in the emerging nations all depend on their going through a similar period of rapid growth.

They are having a hard time doing so: in energy usage, we were there first, and have skimmed the cream of the Earth's easily available energy sources.

From a political and moral viewpoint, we in the developed nations bear a responsibility for the plunder of the past centuries. It is unlikely, though, that a large segment of the population in the advanced countries is going to reduce its standard of living by a substantial amount, voluntarily, in order to share the energy wealth of Earth with the emerging nations. As I will show, there may be an acceptable alternative — a way in which inexpensive, inexhaustible energy sources can be made available to the developing nations without self-denial on our part.

Any technological solutions we employ to solve our problems must, though, retain their logic over a very long time span. As E.F. Schumacher put it:

"Nothing makes sense unless its continuance for a long time can be projected without running into absurdities . . . there cannot be unlimited, generalized growth . . . Ever bigger machines, entailing ever bigger concentrations of economic power and exerting ever greater violence against the environment do not represent progress: they are a denial of wisdom."[4]

These considerations should be in our minds as we examine the technical suggestions contained in this book. I would put them in the form of guiding principles:

1.  A proposal to improve the human condition makes sense only if, in the long term, it has the potential to give all people, whatever their place of birth, access to the energy and materials needed for their progress.

2.  A technical "improvement" is more likely to be beneficial if it reduces rather than increases the concentration of power and control.

3. Improvements are of value if they tend to reduce the scale of cities, industries and economic systems to small size, so that bureaucracies become less important and direct human contact becomes more easy and effective.

4. A worthwhile line of technical development must have a useful lifetime "without running into absurdities" of at least several hundred years.

There are other needs which should, I believe, be met by any development of our industrial society if it is to be successful. It would be desirable if the noise and pollution of our transportation systems could be removed from the environments in which we have our homes and raise our children. Yet we must preserve the freedom of rapid motion even to great distances.

We should also strive for a solution to the problem of unwanted growth in our individual environments. If population growth continues, we should look for a way in which it can do so while still allowing each individual human town to be stable in size and density.

Finally, as we strive to find solutions to the physical problems faced by mankind, we must realize, with humility, that we can offer no panaceas. There are no Utopias. Mankind does not change, and retains always the capacity for evil as well as for good. At the most we can suggest opportunities whose technical imperatives will make it easier for mankind to choose peace rather than war; diversity rather than repression; human simplicity rather than inhuman mechanization. Technology must be our slave, and not the reverse.

Within the past decade four problems have been recognized, all of which relate to the limited size of Earth: they are energy, food, living space and population. The last of these is basic to the other three, therefore we must know the predictions for growth in the human population, and should estimate the accuracy of those predictions. The basic source books for demography are the publications of the United Nations Department of Economic and Social Affairs. There have been four attempts over the past twenty years, by that Department, to summarize worldwide statistics and predict world population growth. The last of these was published in 1973.[5] The resources that were employed for that study are probably at least as great as those available to any other scholarly group.

As a starting point, two numbers are well known: the present world population (just over four billion people, that is four thousand million) and the population growth rate. For several years that last number has averaged 2 percent annually, corresponding to a doubling time of thirty five years for the world population.

Viewed on a time scale of many centuries, though, the population growth rate has itself increased continuously. This has led to such papers as that of Von Hoerner, which shows that up to 1970 the best mathematical fit to the population growth curve would lead to a true "explosion" — an infinite number of people about fifty years from now.[6] This sort of study is of great value in calling attention to the growth problem, but it is best understood as a statement that within the next few decades the growth rate must reduce, and radically. For purposes of this book I will use the much more conservative growth rate figures of the U.N. — the situation is already serious enough without needing to overstate it.

The total world population in 1980 has been estimated by the U.N. Department of Economic and Social Affairs in each of its four summaries, beginning in the early 1950's. Significantly, in each successive revision the Department has raised its estimate of the population of the world in the year 1980. As that date grows closer and the extrapolations can be based on more accurate information, the Department has found that its previous estimates have been too low.

We must also assess the kind of biases that may be put into the Department's numbers as a result of inevitable political pressures. During the past few years many nations have introduced population control measures, enforced either by economic bribery (as in India, where the payment to a young man for undergoing an irreversible vasectomy is typically a quarter of a year's salary) or by social and governmental pressure (as in China, where early marriage is forbidden and where a third child is barred from receiving governmental welfare benefits). When a United Nations member state tells the Department that it has such a program in force, the Department can do little but take the statement at face value. Its predictions, therefore, generally reflect the assumption that the population control programs will be successful as planned. The risks contained in that assumption were illustrated in 1977, when population control became a political issue in India, and the government which had attempted such control was overthrown at the polls after being in office for many years. Even with a successful population control program in the underdeveloped nations, the Department tells us that there will be about six and a half billion people in the year 2000. Growth over this last quarter of the twentieth century within the developed nations will be slow; the increase will come almost entirely within the poor nations. South and East Asia alone will have, by the year 2000, more people than the entire world had in 1970. On the average the one third of the human population that now lives in developed nations is adequately well off in medical care, education, food and material possessions, though many of the larger developed nations have serious problems of internal inequities. By the end of the century, though, an even smaller fraction of all people will live in developed nations, according to the United Nations. The world of 2000, then, will be poorer and hungrier than the world of today.

This growth in population seems contradictory if we recall that the Department's numbers reflect an optimistic assumption about population control programs. There is no contradiction, though — the anticipated increase in numbers will be the result of a distorted age structure of the populations in the poor nations. There, medical advances have come so recently that now most of the population is very young, well below childbearing age. Even if those young people have only two children per couple, the populations of their countries will rise greatly over the next generation.

Knowing that fact, we must also recognize that for the poorer nations not to experience rapid population growth over the next twenty five years, they would have to adopt violent measures. It would not be enough to limit family size to two children — it would be necessary for those nations to suppress new births to levels probably unachievable except by massive, forced sterilization.

The U.N. studies assumed that population growth rates would fall toward the end of the century. The U.N. hardly dares to predict what will happen beyond that time, but if we project their graphs we find that the 10 billion mark will be reached by 2035. Most of the "new people" will be in the underdeveloped nations, and will be born into poverty. And remember — that's the "good" news, based on the idea that population control programs will succeed.

By the same token, as time goes on the U.S. fraction of the total world population will become more and more insignificant. By the turn of the century only one human being in twenty five will be American, and only one in fifty of the new births will be in this country. As far as the total world situation is concerned, therefore, it hardly matters what happens to our own low birth rate.

Though I've used the U.N. figures in estimating how the world population will grow during the next few decades, there are three reasons for being uneasy about doing so: 1) the U.N. figures are based on an assumption that growth rates in the poor nations will be reduced drastically, partly as a result of industrialization. There are, though, serious barriers to the industrial revolution in those nations; 2) the U.N. has been too conservative in its previous estimates, and it may be so again; and 3) to achieve a downturn in population growth rate will mean reversing a trend[7] which has existed for at least 2,000 years. That may not be easy.[8]

In the rich countries, the comfort, the abundance and the freedom of choice enjoyed by most people are achieved only by a high rate of energy use. We grow food efficiently only by spending energy to make chemical fertilizers[9]; our houses are lighted, powered, heated or cooled at the expense of energy; our freedom to travel depends on burning, every year, an amount of fuel which is many times our own weight.

In the United States we now use energy in all its forms at a total rate of about 10,000 watts per person. Until the energy crisis of 1973-74, that use rate was doubling every eight years. Not all of that expenditure of energy is necessary, but our experience during the 1974 gasoline restrictions taught us that not much saving in energy can be made without a noticeable reduction in each individual's freedom of movement. If energy shortages are going to become chronic, we must not forget what they will mean, not only in terms of our inconvenience, but in terms of sheer survival within the poor nations. We must also recognize that conservation is only a palliative — we will continue to need new sources of energy.

At present we in the United States are sharply aware of the need for energy conservation. A number of energy saving schemes have been tried already, but the people whose business it is to anticipate future use of power predict, at best, a lower rate of increase than was common up to 1974.[10,11]

In the U.S. we now burn about half a billion tons of oil products every year, and our total energy usage is about two and a half times as great.[12] A rise in the living standard of the underdeveloped nations to our level would require a similar use of energy.

If the entire population of Earth were to be using energy at the same rate as we do, and were obtaining it from the same mix of oil, coal, gas and other sources, the world total of proven oil resources would be used up in about four years. Even with a strong program of conservation, our use of energy would still be so high that if the whole world were to be at our standard of living, and getting all its energy from oil, by the turn of the century the world use rate would burn up the world's proven resources in only half a year.

There are, of course, large quantities of oil not yet on the "proven reserves" list, but their recovery will probably be at a cost to the environment. In the United States, where the environmental movement began and is strong, there is already concern about the cost to our natural environment of exploiting low grade, remote and undersea sources. For oil, it means the ugliness of the drill rigs in the Santa Barbara Channel and the dangers of the controversial Alaskan pipeline. For coal or oil shale, it means strip mining. For nuclear fuels, it means mining and crushing surface rocks over large areas of the Western mountain landscape.

Inexpensive, abundant sources of energy have been the basis of the industrial revolution so far. Now, when energy costs are rising sharply, those costs may well be contributing to inflation and economic stagnation in the "wealthy" nations. In a single year, 1973-74, the world price of crude oil quadrupled.[13] That single increase cost our U.S. economy more than twenty billion dollars for every year that followed.

In poor, heavily populated countries, rising energy costs are even more serious — in order to grow enough food to lift their increasing populations above the starvation level, those countries must convert very rapidly to intensive agriculture. This conversion will require greatly increased fertilizer production, and that in turn will demand energy.[14]

So far, nuclear power has provided only a small fraction of our energy needs. As fossil fuels grow scarcer and more expensive, most experts think we will be forced to rely more and more heavily on nuclear fuels. The prospect is not an attractive one: the study prepared by Associated Universities, Inc. foresaw most of our electric power coming from liquid metal fast breeder reactors within three decades.[15] The problem of the disposal of their radioactive wastes would not be easy to solve. In addition, these reactors would produce plutonium, from which atomic bombs in large numbers could easily be made. It seems likely that in that case nearly every nation, whatever its size or political stability, would have its stock of nuclear weapons. Large quantities of fissionable materials would be shipped about, and almost inevitably some would also be hijacked by terrorist groups.[16]

For many years we have looked toward nuclear fusion as a clean power source, but even after twenty years of effort and billions of dollars of investment in research no laboratory has succeeded in achieving it. As development has gone on it has also become clear that nuclear fusion will not be so clean a source as originally hoped — it too will produce substantial radioactive wastes. I don't consider fusion power research a waste of time, but it's important to realize that fusion power would require a technology far more difficult, advanced and speculative than anything suggested in this book.

Solar energy would be a good solution to our energy problems, if it was available twenty four hours per day and was never cut off by clouds. We should not dismiss it entirely, but it's very difficult to obtain at Earth's surface when we need it. To summarize, our hopes for improvement of the standard of living in our own country, and for the spread of wealth to underdeveloped nations, depend on our finding a cheap, inexhaustible, universally available energy source. If we continue to care about the environment in which we live, that energy source should be pollution-free and should be obtainable without stripping Earth.

It could be argued that in the most developed countries a slowing of the growth rate in energy usage could occur without serious hardship. That may be true, although I have the uneasy feeling that there may be a connection between energy shortages, price rises, and the present serious economic problems in all the industrialized, energy consuming nations. In the underdeveloped nations, for which the industrial revolution is still to occur, rapid growth rates in energy usage are probably a condition of development. For a healthy world economy it may therefore be necessary to assume that the growth rates which have existed up to now (about 7 percent per year in energy) will have to continue. It has been pointed out by Von Hoerner that if such growth continues, within about eighty five years the power we will be putting into the biosphere will be enough to raise the average temperature of Earth's surface by one degree centigrade.[17] That is enough to cause profound changes in climate, rainfall and the water level of the oceans. Some geologists feel that the ice ages of the past were brought on by temperature changes no larger than that.

I think Von Hoerner is basically right. We can make our own independent estimates, and come up with similar results. Using the "optimistic" low growth rate in population projected by the United Nations, by the year 2060 there will be some 13 billion people. If at that time the present great disparities in the wealth of nations have been reduced, so that all are using energy at about the same per capita rate, that maximum tolerable rate turns out to be greater than our own by an amount that is only 3 percent per year of per capita growth. The "heat limit" is therefore a real one. It may be that it could be pushed back, for a while, by covering large areas of Earth with mirrors to reduce the total of absorbed solar energy. But it cannot be delayed for long — another fifty five years and we would be putting into the biosphere ten percent as much heat as is received by the Sun. A continual growth of energy usage on the surface of Earth, therefore, even if the growth rate is moderate, is one of the "absurdities" of which Schumacher has written.[18]

Professor Robert Heilbroner has studied the consequences, for human political and social development, of the energy and materials limits we have just discussed.[19] He assumes, in my opinion rightly, that people will continue to be guided by the same desires, instincts and fears that have dominated human history up to this point. He dismisses the notion of arresting the industrial revolution at its present level:

> ". . . impassioned polemics against growth are exercises in futility today. Worse, they may even point in the wrong direction . . . In the backward areas, the acute misery that is the potential source of so much international disruption can be remedied only to the extent that rapid improvements are introduced, including . . . health services, education, transportation, fertilizer production and the like."

He is pessimistic about the prospects for widespread social change either within the capitalistic or socialist systems:

> "We have become aware that rationality has its limits with regard to the engineering of social change, and that those limits are much narrower than we had thought . . . that growth does not bring about certain desired ends or arrest certain undesired trends."

In his opinion, as a result of the increasing scarcity of energy and materials,

> ". . . a climate of extreme 'goods hunger' seems likely to result. In such a climate, a large scale reorganization of social shares would have to take place in the worst possible atmosphere, as each person sought to protect his place in a contracting economic world."

Under these conditions Heilbroner feels that the threat of nuclear war is likely to increase greatly in the next decades; because of energy and materials limits,

"...massive human deterioration in the backward areas can be avoided only by a redistribution of the world's output and energies on a scale immensely larger than anything that has hitherto been seriously contemplated ... such an unprecedented international transfer seems impossible to imagine except under some kind of threat.

"Yet two considerations give a new credibility to nuclear terrorism: nuclear weaponry for the first time makes such action possible, and 'wars of redistribution' may be the only way by which the poor nations can hope to remedy their condition."

Even if nuclear war does not occur, and humanity staggers on for another two or three generations, Heilbroner feels that the heat emission limit poses:

"... a challenge of equal magnitude for industrial socialism as for capitalism — the challenge of drastically curtailing, perhaps even dismantling, the mode of production that has been the most cherished achievement of both systems. Moreover, that mode of production must be abandoned in a mere flash of time as historic sequences are measured."

Heilbroner points out that even in the decades immediately ahead we will be forced to turn to increasingly authoritarian governments:

"... the passage through the gauntlet ahead may be possible only under governments capable of rallying obedience far more effectively than would be possible in a democratic setting." "... strong leaders provide a sense of psychological well-being that weak ones do not, so that in moments of crisis and strain demands arise for the exercise of strong-arm rule."

He concludes that intellectual freedom of expression is almost sure to be sacrificed to the exigencies of the energy and materials limits:

"... suppose ... that only an authoritarian, or possibly only a revolutionary, regime will be capable of mounting the immense task of social reorganization needed to escape catastrophe ... might not the people of such a threatened society look upon the 'self-indulgence' of unfettered intellectual expression ... as of no concern, or even of actual disservice, to the vast majority?"

There is, of course, an alternative to industrial growth. Conceivably, perhaps after a series of catastrophes, mankind would adapt a static society. This alternative, a "steady-state" civilization, was considered by J.W. Forrester, leader of the M.I.T. systems analysis team which produced, with the support of the Club of Rome, the document *Limits to Growth*.[20] By calling attention to the consequences of exponential growth in a finite environment, that group performed, in my opinion, a great service. Detailed shortcomings of the computer model used are unimportant by comparison. Forrester could see no viable alternative but a rapid switch-over of our present civilization to a steady-state mode. Heilbroner comes to a similar conclusion:

"In our discovery of 'primitive' cultures, living out their timeless histories, we may have found the single most important object lesson for future man."

A steady-state world order need not be primitive. For example, the pre-Conquest world of the Inca in Peru was a rigidly structured, dictatorial society satisfying a steady-state condition. A peasant of the Inca empire went through life with all his duties and responsibilities rigidly specified, and at his death left a world almost exactly the same as the one he was born into. Almost any static society is forced in self-defense to suppress new ideas. In Heilbroner's words:

"The search for scientific knowledge, the delight in intellectual heresy, the freedom to order one's life as one pleases, are not likely to be easily contained within the tradition oriented static society...."

Professor Heilbroner is frank to admit that:

"...many conclusions in this book have caused great pain to myself ... the human prospect, as I have come to see it, is not one that accords with my own preferences and interests, as best I know them."

And finally,

"If then, by the question 'Is there hope for man?' we ask whether it is possible to meet the challenges of the future without the payment of a fearful price, the answer must be: No, there is no such hope."

# *Chapter 3*
# *The Planetary Hang-up*

The exponential growth of population, on what has become not only a finite but now a sharply limited planet, is almost certain to make the decades ahead on Earth very difficult, and perhaps catastrophic. In the United States, even cushioned as it is by previous wealth, we are feeling the pinch of unemployment, rapid inflation and conflict between industrial efficiency and environmental protection.

If we look in detail at the population growth rates of individual countries, we find that stability has been reached in those areas which have achieved wealth at a high technological level by intense use of energy: North America, Europe, and Japan. To maintain that growth of wealth, these countries must burn fossil fuel reserves at a frightening rate. Between the Persian Gulf and Japan there is a continuous chain of oil tankers, spaced so closely that the crew of one can see the smoke of the next.[1] Our own U.S. appetite for fossil fuels is even greater.

In past centuries plagues and wars were an important factor in holding populations stable. Where poverty is widespread and improvements come slowly, as in South America, Africa, and India, population growth rates remain explosive. Poverty and ignorance go hand in hand, and the decision to limit family size can best be made by families freed of heavy manual labor, secure in good health care for their children, and wealthy enough to spare their children's time from the fields for education.

It seems then that the key to a low population growth rate may be wealth. Conversely, we must be somewhat apprehensive about the accuracy of the U.N. estimates on population growth rate. Those estimates, even though they predict a population three times its present size within one human lifetime from now, are based on the assumption that growth rates in the poor nations will drop sharply long before that. If, though, there is no way that the underdeveloped nations can become significantly wealthier, then the corresponding drop in population growth rates may not come except by catastrophe.

If we want population growth rates to drop by peaceful means, it seems that the best way may be to attack the issues of poverty and ignorance: we need to increase the wealth of the underdeveloped countries not just by a few percent per year, but massively, by factors of ten or a hundred. We can't begin to do this by giveaway programs. We don't possess the enormous amount of wealth needed, and the historical evidence seems to be that small efforts to help are usually canceled by population increases. The areas of the world with the worst problems are often energy poor or located in miserable climates, so their long term prospects for industrialization don't justify optimism.

Somehow we must find a way to bypass those limits, and to set up a chain reaction in the production of new wealth — a reaction that we may trigger but which must then sustain itself as it grows. It will be of little value unless the doubling time for wealth is quick compared to the population doubling time in the poorest areas — and that means short compared to eighteen years.

We are in a period in which technical change comes rapidly. Often the results of change are mixed or heavily adverse, yet we cannot stand still — to do nothing is itself an action, for it is to condemn millions in our crowded world to certain death by starvation. What actions can we take that will reverse the present trend toward increasing poverty and hunger?

Several years ago Gerald Feinberg addressed the issue of technical change in a book subtitled *Mankind's Search for Long-Range Goals*.[2] I would take issue with Feinberg on only two points: we don't usually "search" for goals; most people have enough to do to cope with their own lives, and leave the long-range or large-scale questions to chance, with perhaps a dim hope that "something will turn up." Second, it was Feinberg's suggestion that major issues be submitted to as large a fraction of the world's population as possible. He had in mind particularly such potentially explosive issues as artificial genetic change, the alteration of personality by chemicals, and increased human longevity. The idea that large issues should be debated by many people, not just a power elite, is in my opinion a good one — in the United States during the past decade it has been put into practice with good effect by voluntary citizen movements in the areas of family planning, environmental protection and land conservation. We must recognize, though, that a population must be relatively wealthy, well educated, and have considerable leisure if it is to spare the time and effort for such debates. In the areas of our world where the problems are most severe, almost no one can spare the effort to think beyond the next meal.

This is one of the rare occasions in human history in which a new technological option is being subjected, deliberately, to wide popular debate before, not after, the decision to go ahead with it has been made. I prefer it that way. I believe that the concept of the humanization of space can stand on its own merits, survive detailed numerical checks, and survive logical debate. To support it requires no act of faith, only the willingness to study unfamiliar ideas with an open mind. In keeping with Feinberg's strictures, in my opinion the long-term goals we should set, relevant to space habitation, should only be those with which nearly every rational human being, possessed of goodwill toward others, could agree. I think that the following goals satisfy that criterion, and that they should be our most important goals, not only for humanitarian reasons, but for our own self-interest; and I do not believe that those two justifications must necessarily be in conflict.

1. Ending hunger and poverty for all human beings.

2. Finding high quality living space for a world population which will double within forty years, and triple within another thirty, even if optimistic estimates of low growth rate are realized.

3. Achieving population control without war, famine, dictatorship or coercion.

4. Increasing individual freedom and the range of options available to every human being.

We in this country are certainly going to become increasingly insignificant as the years go on, both because of our decreasing relative numbers (only 4 percent of the world's population by the year 2000) and because of the energy and materials limits to the growth of our wealth. Is it reasonable then to set a fifth, more parochial goal? Realizing our limitations, should we not seek a role for this country that can be of benefit to humanity as a whole, and can at the same time benefit directly our own people and our own economy?

Considering the first four goals in the context of the fifth, it should be clear to us that we have no special magic to export in regard to governmental systems. Most of us are passionately attached to a democratic form of government, but it doesn't travel well — while much of the world has rushed to imitate our technology and our systems of productivity, at the same time, there has been no such rush to imitate our system of government. We must also recognize that other systems have been found to work, perhaps not very much worse than our own, even in societies that have achieved industrialization. I own to a private belief that wealth and leisure, shared generally by a large segment of a population, are powerful forces tending toward more democratic forms of government, but I suspect that if the human race does achieve general affluence, and with it an increase in real human freedoms, it will do so within the outward forms of many different forms of government, and with many of the old polemic catch phrases still in regular use.

Can we point the way to an exponential growth of wealth, which could continue for many centuries, and which could be shared by all people? If we can, and can further lead the way to it by virtue of techniques in which we are acknowledged the leaders, we will have done something, as a nation, very worthwhile — indeed, something far more worth looking back on with pride than lost dominance or a vanished empire. To achieve such an exponential growth of wealth, and therefore the opportunity to reach the four great goals listed, we would need:

1. Unlimited low cost energy, available to everyone rather than just to those nations favored with large reserves of fossil or nuclear fuels.

2.  Unlimited new lands, to provide living space of higher quality than that now possessed by most of the human race.

3.  An unlimited materials source, available without stealing, or killing or polluting.

Nothing in our solar system is truly unlimited, of course; no expansion can go on forever. But an exponential growth of wealth can be considered rationally if we can find the environment in which that growth can proceed for many hundreds of years. There is an enormous difference between sharp limits, forced on us within years or decades at a time when most of us are still in deep poverty, and limits reached only after several hundreds or thousands of years, under conditions of high prosperity and universal education in a generally affluent and literate human population.

We are so used to living on a planetary surface that it is a wrench for us even to consider continuing our normal human activities in another location. If, however, the human race has now reached the technical capability to carry on some of its industrial activities in space, we should indulge in the mental exercise of "comparative planetology." We should ask, critically and with appeal to the numbers, whether the best site for a growing advancing industrial society is Earth, the Moon, Mars, some other planet, or somewhere else entirely. Surprisingly, the answer will be inescapable — the best site is "somewhere else entirely."

In a roundtable TV interview, Isaac Asimov and I were asked why science fiction writers have, almost without exception, failed to point us toward that development. Dr. Asimov's reply was a phrase he has now become fond of using: "Planetary Chauvinism."

What do we need for the exponential growth of wealth? Three things: energy, land area and materials. The next question is: How much of them will we need, if growth of any sort is to go on? Suppose that a universally affluent, energy rich, educated human population has about as low a growth rate as would be noticeably different from zero — an increase of the total human population by about one sixth over a human lifetime. That very modest rate of increase, considerably less even than now exists in the developed nations on Earth, would result in total growth by a factor of 20,000 over a period of 5,000 years. Right now, of course, growth is ten times as fast.

The conclusion we have to draw from those facts is that for exponential growth of wealth over a time span long enough to make a real qualitative difference in human history, the factors needed in energy, land area, and materials are not just two or four or even ten: they're at least in the thousands, and probably in the hundreds of thousands. It is with that in mind that we must evaluate Earth and its "competitors" as sites for a large industrialized civilization.

The energy limits on planet Earth were discussed in the first chapter. Even if some inexhaustible source of energy is discovered and exploited here, we will reach the heat barrier in about one and a half human lifetimes. We cannot base an expanding industrial civilization on a site where a fundamental limit will be reached that soon.

The land area resources of Earth are known. Its geometry as a sphere in space determines that some of its areas are heated moderately, others too much or too little.

We could, in principle, make all the land area of Earth habitable, including Antarctica, and we could float colonies on the oceans. The resulting changes in worldwide climate would be profound, and there would be a severe risk of melting the ice caps and precipitating another ice age, but we would be forced in that direction if we had no alternative. The era in which virgin land of good quality in a good climate was available for settlement is long past. The United States is a relatively uncrowded country by world standards, but already our fastest growth is in regions (Arizona, New Mexico, and other desert areas) which would not attract large numbers of people if there were no air conditioning. In areas of California, once regarded as highly desirable, overcrowding has become so bad that in a recent survey about a third of the Californians said they'd rather be living in some other state. But the mood in nearby states of low population density (Oregon, Idaho, and others) is openly hostile to emigrants from California. In Europe, the Netherlands is already near the saturation point for population, given its climate and growing season. In much of Asia the crowding of the land is still more serious, and it is there that the great population increases are still to come.

The prospects for colonization of other planetary surfaces are unappealing. First, the total areas involved are too small — the Moon and Mars total only about the land area of Earth, and neither has an atmosphere. Both have the wrong gravity for maintenance of our bodies in good health, and the Moon has a fourteen day night, which would require any colonists there to do without natural sunshine for weeks at a time. Venus is an inferno hot enough to melt some metals, and would be uninhabitable without extensive "terraforming" of a kind well beyond our present capabilities. Even after such a conversion, it would still be unbearably hot, because of its location so much nearer the Sun than is our Earth. Finally, the total area of Venus, about equal to that of Earth, would be worth only two or three decades of time in terms of present growth rates.

Travel away from a planetary surface requires high thrusts and precise timing, and is therefore relatively difficult and expensive. On a planetary surface we are the "gravitationally disadvantaged," at the bottom of a deep potential energy hole. From Earth, to raise ourselves into free space is equivalent in energy to climbing out of a hole 4,000 miles deep, a distance more than six hundred times the height of Mt. Everest. Does it make sense to climb with great effort out of one such hole, drift across a region rich in energy and materials, and then laboriously climb back down again into another hole, where both energy and matter are more difficult to get and to use?

There are still other disadvantages of basing an industrial civilization on a planetary surface:

Solar power: On Earth it is attenuated by the atmosphere, uncertain due to weather, and cut off every night by Earth's rotation. The average of solar energy input[3] to the United States, over a year, is only about 0.18 kilowatts/m². In free space at a distance no farther than the Moon, but well away from both Earth and Moon, solar energy is available full time at a rate of 1.4 kilowatts/m² — that is, almost ten times higher than at Earth's surface, when averaged over a year, and it's never cut off by night.

Travel and shipping: On a planet with atmosphere both are slow, and wasteful in energy. In the U.S. transportation system, about a quarter of all the energy we spend goes into fighting gravity and atmospheric drag — that's a waste of around two and a half tons of petroleum each year for every man, woman and child in our country.

Confinement to one gravity: Until the last decade, it would never have occurred to us to think that industry could operate in zero gravity, but if that option is presented to us, it can be used to good advantage. Every activity involving massive objects or large weights of materials is dominated by the cranes, rails, engines and other machinery needed to handle heavy objects in Earth-normal gravity. In zero-G, all that would be unnecessary. There are industrial processes, such as the growth of perfect large single crystals, which are impossible in one-G, but easy in zero-G. Single crystals can be ten or twenty times as strong for their size as the same materials in less ordered form.

Climate, the locations of materials, and the special property of oceans for cheap transportation tend to cause wide separations, here on Earth, between agricultural producing areas and population centers. As a consequence, we become tied into interdependent networks thousands of miles in extent. Anyone who interrupts one of those networks by cutting off our sources of energy, of food, or of materials can hold a large population for ransom. We have examples of that sort of threat frequently, and the result is always the same: even at best, prices are driven up, production is slowed down, and almost everyone suffers. At worst — and that worst is approaching with frightening speed as we plunge deeper into the energy-food crisis — we approach a world society governed by mutual threat: deprive me of oil, and I deprive you of food; threaten me enough with deprivation, and when I have nothing left to lose I will risk life itself in a last desperate gamble; provide for me, or I burn you to death with hydrogen bombs.

The same factors of climatic variation, the need for sea transport to minimize transport inefficiencies caused by gravity, and the seasonal cycle tend to produce very large concentrations of population — living in numbers so great that they are continuously subjected to the evils of bigness: high crime rates, dirt and disease, social alienation, and political corruption.

Up to now, we have taken it for granted that huge cities were an inevitable part of industrialization. But what if it were possible to arrange an environment in which agricultural products could be grown with high efficiency, anywhere, at all times of the year? An environment in which energy would be universally available, in unlimited quantities, at all times? In which transport would be as easy and cheap as ocean freight, not just to particular points but to

Solar power is almost ten times higher in space than on Earth and it's never cut off by night. Solar energy converted to microwave frequencies can be beamed to Earth.

everywhere? There is, now, a possibility of designing such an environment, and it will be the topic of the next chapter.

We have a clear decrease in human options. The solution to the problems of energy and materials would not guarantee freedom and well-being for all — we've had too many examples in history of man's capability for inhumanity for us to assume that. Until very recently, though, we had some hope that averaging over the ups and downs the human race as a whole was struggling toward more decent living conditions, better education, and more freedom. The ignorance and cruelty of a Genghis Khan, the sadistic mad genius of a Hitler, were, we hoped, temporary horrors in a slow development averaging toward the better. But while we remain limited to the surface of a gradually depleted Earth, we face a new kind of threat, and even our success becomes failure. Survival will require that either voluntarily or under coercion we must limit our options. Heilbroner has argued that those limits will almost surely be more than physical, and that in the long run the freedom of the human mind will have to be limited also, as it is, very severely, in primitive human societies which have achieved enforced stasis through a rigid social code.

We are surely far from having found the best ways in which human beings can live together and govern themselves; and surely far from having achieved freedom for all, or having explored all the talents of which the human mind is capable. What chance will we have, though, here on an Earth ever more crowded and more hungry for energy and materials, to allow for diversity, for experiment, for groups to try in isolation to find better lifestyles? What chance for rare, talented individuals to create their own small worlds of home and family, as was so easy a century ago in our America as it expanded into a new frontier? For me, the age-old dreams of improvement, of change, of greater human freedom are the most poignant of all. And the most chilling prospect that I see for a planet-bound human race is that many of those dreams would be forever cut off for us.

# Chapter 4
# New Habitats For Humanity

Biologists and botanists talk of the "range" of a species — the limits, on the surface of Earth, over which a species can survive, grow and reproduce. For our ancestors of the remote past, the range was the tropical ocean. It was a major step in the development of living beings when the early amphibians evolved into air breathers. Now, when we are about to design new habitats for man, we must question what limits are set by our own physiology. As we ask those questions, we must be conservative in our answers — we're not asking for extremes: not for the limits that apply to highly motivated athletes in superb physical condition, to mountain climbers, astronauts, or deep sea divers, but for those that apply to quite ordinary people — ultimately, to "Aunt Minnie in her rocking chair." That conservative approach should apply even to the first habitat we build, for a practical reason that has a basis in hard economics: when people are called upon to work under hardship conditions, in miserable climates or exposed to disease, they have every reason to leave their families at home, and to demand high pay for their hardships and deprivations. Pay scales on the Alaskan pipeline construction job have to be very high. Even our first space colonies must pay their way, and they can only do so if they do not price themselves out of their markets. They must be places to which people will come by choice, and to which their families will enjoy coming also — places where it will be possible to live and work and raise children in ease and comfort.

With this conservative approach, we must then ask what constitutes a human environment — what is the "range" of mankind as a species? Most of us are accustomed to living near sea level. A large fraction of humanity, though, in mountainous areas of every continent, lives at altitudes as high as Denver, Colorado, where the pressure is 20 percent lower; and that fraction includes people who are elderly — a slightly lowered pressure doesn't seem to bother them.

The Federal Aviation Agency, to assure that pilots will be in a state of full alertness, requires that oxygen be used for any flight above 12,500 feet lasting more than a half hour. As a sailplane pilot, with my oxygen mask always at hand, I like to take a few breaths of oxygen at the tops of Western thermals, which are often a good deal higher than that. Serious mountaineers, climbing by muscle power and carrying packs, go far higher without oxygen, some to as much as 25,000 feet. Few human habitations, though, are more than twice as high as Denver, and in those few, within the Andes and the Himalayas, the population has adapted through natural selection over many generations to life at low pressure. In space habitat regions where people may be called on to do very light work not lasting more than a few hours, we can take the Federal Aviation Agency's limit as a guide, and for conservatism we should probably maintain in space habitat living areas an oxygen pressure at least as rich as Denver's, a mile high.

As deep sea divers and astronauts have shown, the nitrogen that makes up most of the atmosphere is unused by our bodies. On Earth, nitrogen serves to inhibit flames, and acts as part of our cosmic ray shield, but we do not consume it except through the food we eat. Curiously, neither do many plants: they take up nitrogen through their roots, from the soil, rather than from the air. If we provide some alternative way to inhibit flame and to protect ourselves from cosmic rays, the range of the human atmospheric environment will go as far as an oxygen atmosphere with the same oxygen pressure that is found in Denver. Though astronauts have lived in such atmospheres for several days while on the lunar surface, long-term tests with larger numbers of people will be needed before we can be sure that no respiratory problems will develop.

First we have considered air, the medium without which we would be dead in a few minutes. Next we can think of the range of temperature and climate over which humans can live and work. That range is wide, from the deep freeze of the "Pole of Cold" in Siberia to the heat of the Sahara in midsummer. The range of comfort, and of easy operation without heavy clothing, is much narrower — just the few degrees where we set our room thermostats when we have the choice. Outside that range our efficiency goes down, and the steady migration to regions of mild climate without great variations suggests that the human desire for a comfortable temperature runs deep. We'd better plan on a narrow temperature range for most human activities, but allow for the variations needed for sports like skiing.

With atmosphere and mild climate, we can survive for one or two days. Without water, though, we can't last much longer than that. Nearly all of the mass of our bodies is water, and in desert areas the inhabitants seldom deal with more than a few pounds of extra water per person. We're looking toward a pleasant, not a parched environment however, so we'll be much more generous — for the moment we will think in terms of several tons of water per person.

In extreme conditions people can go for several weeks without food, but in the space communities there will be no difficulty in providing food of a richer abundance, and with greater reliability, than exists over most of Earth. Water and food are not limits on the range of the human species in space.

Zero gravity requires acclimatization, and for some people the adjustment takes several days. All three men of one Skylab crew were ill during the first twenty four hours. Skylab tested a small sample of very healthy human beings for 90 days, and during that time their bodies underwent definite physiological changes: a loss of blood volume, degeneration of certain bones, loss of bone marrow, and a slackening of muscle tone. Those changes were reversed and recovery was complete after some weeks when the men returned to Earth, but the advisability of exposing people to zero gravity for many months without change seems doubtful. It's likely that a heart which has grown used to the easy conditions of zero gravity might be prone to failure when gravity is restored. We don't want to make emigration into space a one-way trip, without the option of return at will.

Curiously, we all have the experience of what amounts to zero gravity every day of our lives. Physiologists have found that bed rest takes the load off the body at least to the same extent as does zero gravity, and that all the same types of degenerative changes occur in the two cases. We know that it is not necessary to be subjected to one gravity all the time — a few hours each day may be quite enough. How much less, we don't yet know, but it seems wise to plan that the areas where people will spend their time when they're not working will be at approximately Earth-normal gravity. Ordinary people won't put up with the Skylab substitute for it, which was an intense program of exercise occupying more than an hour every day. Fortunately again, gravity is easy to find in space: rotation can provide it. On the inside of a hollow, rotating vessel the gravity can be made to be the same as on Earth, and if the vessel is big enough the human body will find the artificial gravity indistinguishable from the real thing.

On Earth, sensitive, delicate organs within the inner ear have evolved to measure changes In the position of our bodies. Although they have their limitations, these organs can detect rotation about any of three axes.

Within a rotating environment, with a rotation period measured in fractions of a minute rather than twenty four hours, our motion sensors can detect the fact that "all is not normal" as far as gravity is concerned. For a number of years, physiologists have conducted studies to find out how difficult it will be for people to adjust to a rotating environment. The principal centers for these studies have been the U.S. Naval Medical Center at Pensacola, Florida and the Soviet space program's ORBIT centrifuge facility in the U.S.S.R. Although there are limitations to the completeness with which such Earthbound tests truly duplicate conditions in space, there appears to be general agreement on the following points: first, almost no one has any difficulty in adjusting to rotation rates of one per minute or less. Second, as the rate climbs above two, three, four rotations per minute and even higher, more and more people find it difficult to adjust — they experience a variety of unpleasant symptoms ranging from motion sickness to drowsiness and depression. Some, though, are able to adapt to rotation rates as high as ten rotations per minute. In the case of a habitat in space, the range of interest is between one and three rotations per minute — high enough to be of concern, but low enough that most of the subjects so far tested have been able to adapt to it, usually within a day or two. For the larger habitats, which will almost surely follow the first small "models," the rotation rates can be kept below one rotation per minute without compromising efficiency of design. For the earliest habitats, economy appears to dictate that a rotation rate of about two RPM be chosen, for Earth-normal gravity, and that the applicants for jobs in the early habitats undergo tests to determine whether they are unusually vulnerable to motion sickness in space. The evidence from United States and Soviet space programs so far is that there is very little correlation between motion sickness as we encounter it in aircraft and boats, and the sort of "space sickness" that may be found when we substitute rotation for natural gravity. On the basis of the tests at Pensacola and in Russia, we can guess that only a few percent of the applicants for positions in the early habitats may find, after a few days or weeks in a low-orbital space station, that they are unsuited to life in space.

We have talked of the necessities of life, but if we are to work and live in space by choice, and enjoy doing so, we will ask for more: the age-old human desires of comfort, good food to eat and good wine to drink, room to stretch our legs, good places to swim and to get a suntan, and variety in travel and amusement. We humans have definite ideas about our needs for enjoyment and amusement, and any successful space community will have to accommodate them.

We evolved as a hunting / gathering species, in the light of the sun, and our bodies need some exposure to it for well-being. Without sunshine, children develop rickets, and without sunshine people tend to grow moody and depressed. Almost surely, the high suicide rate characteristic of the Scandinavian nations is, at least in part, connected to cloudy skies and long cold winters. A successful space habitat will have to admit natural sunshine, and that should not be hard to arrange. In space, remote from any planetary surface, full sunshine is available whenever we want it. But to avoid throwing off our internal biological clocks, evolved in a twenty four hour day, we will need to provide a day / night cycle.

When humans existed in small bands, they camped and always stayed near clear running water. Except for their own smoky campfires, the air they breathed was clean. In our pollution-ridden world, no longer can we take clean air and water for granted — most large rivers are dirty. In a space habitat we should make a fresh start, and set up our industry and economy to keep the air and water clean.

Our Earth is rich in plants and animals, but as industry and the human population crowd environments it is not as rich as it once was. City children become starved for the sight of a tree, and in desert areas the palms of the oases have an importance no dweller in a lush climate can imagine. For our psychological well-being, as well as for the cycling of the oxygen we breathe, we should have grass, trees and flowers. Many animal species are a pleasure to us, and if we move into space both we and they will benefit by our taking them along — perhaps, like Noah's passenger list, two by two. Along with the domestic animals, we will certainly want to bring squirrels, deer, otter, and many others. And birds, and some types of harmless insects for them to eat. In space, though, we have an option that doesn't exist to us on Earth — to take along those species which we want and which form parts of a complete ecological chain, but to leave behind some parasitic types. How delightful would be a summertime world of forests without mosquitoes! Perhaps, too, we can find less annoying scavengers than the housefly, and can take along the useful bees while leaving wasps and hornets behind.

Perhaps because we were originally a hunting and gathering species, the urge to travel and to seek out variety in habitat and environment is deeply rooted in many of us. Now that long distance jet travel has become commonplace, a large segment of the population in the developed nations travels regularly for vacations. Our young people are learning wide horizons at a much younger age than did their parents. Some of the results are unattractive — such as traveling drifters, subsisting on doles from home and roaming the world as what the East bloc nations call parasites — but if we believe in humanity we must also believe that the widening of horizons and the interaction of different lifestyles is, on the whole, a good thing, that it tends to cut away the hostilities and the myths that go with isolation, and so tends to reduce the likelihood of wars. Freedom to travel is precious, and adds greatly to human options. Its blockage by poverty or by dictatorial governments always constitutes a loss. We can be grateful, then, that the technical imperatives of the humanization of space are toward easy travel at low cost. We cannot prevent the occasional abrogation of that freedom by a suspicious or reactionary government, but we can at least make sure that no barriers of poverty or energy shortage act to prevent travel.

The growing of food is the most vital of all our industries, and now that we are freed of the planetary hang-up we must ask: What are the optimum conditions for agriculture?

An adequate source of clean fresh water must always be at hand. In a space habitat, water once introduced can be recycled indefinitely, given an inexhaustible source of cheap energy.

The uncertainty of the Earthbound climate is the great bane of all farmers — drought, frost or long continued cloudy weather can ruin crops. Worse still, farming has always been subject to the cycle of boom-and-bust: in a good year, every farmer grows too much, and prices drop for his produce, and in a bad year, he has little to sell although prices are high, and the consumer must pay highly for poor quality. In a space habitat, although people may want to live in climates that vary widely, crops should be grown in constant conditions, dependably unchanging from year to year.

Throughout most of the world only a part of the year is suitable for growing, and when winter strikes it stops all farming over a distance of thousands of miles. If we have a choice, we should provide that agricultural areas, in close proximity to each other and to the consumer, have the seasons and seasonal variations that are best for their particular crops. To make sure that our tables have fresh vegetables and fruit in all seasons, our growing areas should be staggered in phase — January in one while there is June in another. Impossible though that is on Earth, it will be easy in space.

On Earth, all of our high yield grains, and all of our fruits and vegetables, are subject to attack from various pests and viruses. Usually, these pests have evolved through centuries to attack certain plants, and on Earth winds and human travel threaten always to spread plant diseases to new areas. In space, it makes sense to start our agriculture with carefully inspected, pest-free seeds, and to introduce only those bacteria essential for plant growth. If our agricultural areas are separated from our living areas by even a few miles, and receive only sterile water and chemical fertilizers, the vacuum of space will serve as a perfect barrier to keep them pest-free. For the first time, we will be able to have agriculture of high yield without pesticides, insecticides, or crop losses due to raiding birds and animals.

As agriculture has become more and more sophisticated, it has become ever more factory-like. In modern high yield agriculture, the soil in which crops are grown is relatively unimportant; it serves only as a matrix to hold the growing plants. The highest yields are obtained by intense application of chemical fertilizers, and by careful control of trace elements and the acidity of the soil. As the evolution from a pastoral economy to an agri-

Agricultural areas in space can have staggered, controlled seasonal variations so that fresh vegetables and fruit are available year round.

cultural industry has gone on, that industry has become continually more energy intensive. The cost of fertilizer production is dominated by the cost of energy.[1] In space a method for the production of fertilizer will become easy, though on Earth it is uneconomical — the simple heating of an oxygen-nitrogen mixture, in a tube at the focus of an aluminum foil mirror in sunlight, to a white hot temperature. At that heat about 2 percent of the molecules will dissociate and recombine to form nitric oxide, an energy rich precursor of chemical fertilizer.

It appears, therefore, that space can provide the ideal conditions for a highly efficient, totally recycling agriculture, no longer at the mercy of weather and climate.

We are examining the needs of an industrial civilization, so we must look toward the conditions in which industry can work efficiently, at low cost, and free of pollution.

Industry is energy intensive, and with increasing sophistication and the continuation of the industrial revolution, that hunger for energy also grows. Here on Earth, where our energy sources are limited, we have come to think of intense energy usage as very nearly immoral. But if we have a truly unlimited energy source, there is no reason to curtail the natural development of the industrial revolution.

Industry uses energy in two forms: electrical and thermal. Thermal energy is used for melting metals, for raising chemicals to temperatures at which they react, and for making ceramics. On Earth most of the fossil fuels that industry uses are burned to provide this thermal energy. In zero gravity, far from a planet, the concentration of the unvarying, intense sunlight of space by very lightweight, inexpensive mirrors can provide all the energy that industry will ever need. A simple reflector the size of a football field, weighing no more than a car, when extended in space can provide a great deal of process heating. To equal it, an Earthbound factory would have to burn a million barrels of oil every thirty years — but the reflector in space will go on supplying that same power at no cost, as long as the Sun shines.[2]

I spoke of the ease of obtaining electric power in large quantities from sunlight in space. We can be more quantitative about it: given an industry in space, at which large turbogenerator power plants can be built, we can expect to build them for about the price of a coal-fired plant on Earth.

The space power plant, running at zero gravity, will need less maintenance than its Earthbound counterpart; even though its turbine rotor and generator armature may have a mass of thousands of tons, they will weigh nothing in zero G, and can be supported, with no direct frictional contact, on air or magnetic bearings which should have an infinite lifetime. The fuel cost for a plant in space will be zero, so the entire cost of power will be that of amortization, maintenance and distribution. In space the industries that use electric power can locate anywhere in a volume, rather than on a flat surface, so they can be much closer to the power plant, reducing distribution costs. Maintenance should be low, because there will be no fuel handling machinery to service and no friction bearings to wear out.

Putting all the numbers together, a turbogenerator plant running on solar energy in space should be able to supply electricity to nearby industries at a fraction of a cent per kilowatt-hour. That figure is lower than the cost of electricity in all parts of the U.S. except where hydroelectric power is available. After amortization, costs should drop to those of maintenance. The cost of power enters into every part of an industrial economy, so in space it should be possible to produce most goods more cheaply than on Earth.

There is an additional component to energy costs, a component whose force we are starting to appreciate — the cost of uncertainty. When the planners of a new industry cannot predict how much electric and thermal power is going to cost at the time a new facility will be finished, they find it very difficult to make the decision to build, and even more difficult to persuade a lending agency to advance the money for construction. In space, that uncertainty will be removed, because fuel costs will be zero and can be guaranteed to remain so for the life of the Sun — several billion years at the best estimate. Lloyds of London should be very willing to insure a new industry against its power costs going up with that kind of backing!

We should examine whether nuclear fission or fusion power on Earth can ever equal the low costs of solar power in a space colony. The answer seems to be, No, Earthbound nuclear power will not compete successfully with solar power in space. First, for all process heating needs, in space a simple mirror with no moving parts, located at the point of use, will be sufficient. On Earth one would have to go through the expensive and inefficient intermediate step of converting from nuclear power to electricity and then back to heat, since nuclear plants cannot be made in small sizes. For electric power on Earth from fusion, we will overlook for a moment the fact that billions of dollars and twenty years of effort have so far failed to make nuclear fusion a practical reality. Even if it succeeds, its cost will almost surely be much higher than those of a solar plant in space. In a fusion plant, one will first have to spend energy to separate the one part in 5,000 of heavy water from ordinary water, then obtain deuterium from it. Then it will be necessary to pass through a stage of complicated, high technology machinery, involving either lasers or giant magnets. In the end, one will have heat — only to put it into the boiler of a turbogenerator plant. The space borne solar plant will bypass all the hard part of this complicated sequence because it will begin with free solar energy. Finally, the distribution costs in space will be far lower, because distances from power plant to industry will be only a few miles, and because solar electric plants, unlike nuclear stations, can be made in small, convenient sizes adjacent to heavy power users.

In addition to the advantages of zero gravity for the handling of massive objects, for the heating of materials to high temperatures without the contamination of confining crucible walls, for the formation of uniform mixtures of heavy and light materials[3] and for the growing of large single crystals, industry in space will have an additional degree of freedom. By gentle rotation, it will be possible to maintain very thin surfaces accurately in the form of cylinders and cones. That may be especially useful in the case of large mirrors made of thin foil.

Here on Earth our lowest cost transportation is that of crude oil in supertankers. Though the rates fluctuate wildly, tanker construction being about as speculative as pulling the handle on a Las Vegas slot machine, the bare operating costs amount to about 0.06 cents per ton-mile.[4] For shipment of commodities in bulk from one space colony to another, at a speed typical of highway driving on Earth, a tanker-size payload can simply be put in one large motorless container, and accelerated by an electric motor and cable to its drift speed. No crew need go with it, because in the vacuum of space its trajectory and its time of arrival will be known exactly, and there will be no weather or navigational hazards to contend with. The energy cost of such a shipment will be absurdly small — only about a thousandth of the cost per ton-mile that a supertanker works for on Earth.

Commuting to work from a space colony should be correspondingly easy and inexpensive. The typical vehicle can be a sphere, protected from cosmic rays by a dense, foot thick outer shell. It may contain seating on three levels, and be entered by three airliner-type doors. With a comfortably generous amount of elbowroom and leg room for each passenger, about like those of first class seating on a long distance airliner, the sphere can accommodate a hundred passengers. In less than a half minute, an electric motor and cable can accelerate the sphere to the speed of a jet plane, and the flight to a factory a hundred miles or so from the colony will take only a few minutes of vibrationless flight. Just time enough to skim the morning news, and an arresting cable will slow the sphere to its destination. The energy cost? Less than fifty cents per passenger.

Each time the balance is tipped for a particular industry, so that production in space becomes cheaper than on Earth, we will be relieving Earth in two ways: we will be removing the burden of energy usage and materials mining for that industry, and we will be generating an additional force to draw away population — the work force of that industry, and the families of the work force. For many years, the only industries in space that will compete directly with those on Earth will be industries that require no material shipment of material products back to Earth. There are at least two of these: fabrication shops to produce satellite solar power

One of the first large-scale space industries will be solar power satellites to relieve the burden on the world's dwindling energy sources and associated ecological damage.

stations, for location in geosynchronous orbit above a fixed point on Earth's surface to beam down power for Earth's electric systems; and assembly plants for the aerospace industry, building ships for transport among the colonies and from Earth out to the colonies.

For energy in the United States alone, we now burn literally billions of tons of irreplaceable fossil fuels every year. From a conservation viewpoint, it makes little sense to blow away this oil and coal in the form of smoke. It should probably be conserved for use in making plastics and fabrics. That environmental consideration, reinforced by a powerful economic drive, suggests the construction of solar power stations for Earth as perhaps the first major industry for the space colonies.

Within the colonies themselves, no conflict need ever arise between using carbonaceous materials for energy and using them as they should be used: for the petrochemical industry. As we have seen, the cost of solar power in a space colony will be so low that it will be ridiculous there to obtain energy in any other way.

For the continued growth of wealth, a developing economy must have an assured source of materials. On Earth, we are already forced to work poorer sources to obtain our metals. For iron in the United States we have long since depleted the Mesabi range in upper Michigan. As we work poorer veins, the conflict of mining with the environment rapidly becomes more serious. When the ore content is only a tenth of that in a rich vein, we must mine and process ten times as much material to get the same quantity of the metal we seek.

Raw materials mined from the Moon can provide most of the necessary resources for space manufacturing and construction.

In space, our first mines will almost surely be on the Moon. Particularly on the lunar Farside, enormous quantities of materials could be removed without ill effects of any kind. It comes as a surprise to most people to learn how rich a source of industrial materials the Moon is. I believe that in the long run the Apollo Project, much criticized as it was during its lifetime, will be seen to have been of enormous value for its lunar prospecting function. A typical Apollo sample contains, by weight, more than 20 percent silicon, more than 12 percent aluminum, 4 percent iron, and 3 percent magnesium. Many of the Apollo samples contained more than 6 percent titanium by weight. Titanium is in great demand as a strong, light metal, which holds its strength up to a very high temperature. Its present use is mainly in the aerospace industry. Processing it requires high vacuum, high temperature, and a lot of energy — all things which are expensive on Earth but will be cheap in space. Finally, the lunar surface is more than 40 percent oxygen by weight. Strange to think that such a lifeless, sterile landscape contains, locked in its soils and rocks and waiting to be used, the one element we need most to sustain our lives.

In the "long" run, within one or two decades after the human use of space begins, we will begin to exploit the resources of the asteroid belt. For transport in space we must think in terms of energy rather than distance, because travel in space is without atmospheric drag. To bring a ton of material, efficiently, from Earth's surface to the site of a space community would cost about the same, in energy, as to bring that ton of material to the same point from the asteroid belt. The difference is that lifting it from the surface of Earth requires a rocket able to supply more than a ton of thrust, and further requires elaborate fast acting control systems operating with split-second precision. By contrast, moving a load of freight from the asteroids to the colonies can be done in a leisurely fashion, with efficient, low thrust engines. If the engines break down, there will always be plenty of time to fix them, just as a freighter on Earth's oceans can lie dead in the water for days if need be while its engines are being repaired.

Bringing materials from the lunar surface to the site of the space communities will be even easier. The energy cost per ton will be only about one twentieth as much as for shipment from Earth or from the asteroids. As we shall see in later chapters, materials can be brought from the Moon at an initial cost of only a few dollars per kilogram. Later, when space borne industry is well established, the ultimate costs should drop to only a few cents per kilogram.

The Moon is poor in three elements that we need for life and for a full industrial base: hydrogen, nitrogen, and carbon. Apparently, during its lifetime the Moon was subjected to baking at a very high temperature. Fortunately, it has been shown by analyzing the spectra of sunlight reflected from asteroids that some of them are rich in carbon, nitrogen and hydrogen — they are about as good a source for petrochemicals as oil shale.[5] Corroborating evidence for the presence of these elements in the asteroids comes from about twenty meteoroids found on Earth's surface; of a type called "carbonaceous chondritic." The normal economic decisions that govern industrial operations will therefore probably lead to mining the lunar surface for most elements, and the asteroids only for the materials which the Moon lacks. Long before an appreciable fraction of the lunar surface has been mined, it will become easiest to obtain all the materials for colony construction at the asteroids themselves.

Although the total volume of the asteroids is far smaller than Earth's, it is a volume much more accessible than the depths of our planet. On Earth only a thin skin of material is available to us without deep mining under high pressures and intense heat. Even if we were to excavate the entire land area of Earth to a depth of a half mile, and to honeycomb the terrain to remove a tenth of all its total volume, we would obtain only 1 percent of the materials contained in just the three largest asteroids. A striking contrast: we would have to disfigure the entire Earth to obtain only a hundredth of the material contained in three now useless, lifeless asteroids — and there are thousands of those minor planets. Moreover, to bring material into space even from the biggest asteroids requires climbing a gravitational hill only five to ten miles high, instead of Earth's 4,000 miles.

As a reader of science fiction in childhood, I gained no clue that the future of mankind lay in open space rather than on a planetary surface. Later, when logic and calculation forced me to that conclusion, I searched for evidence that others before me had come to the same realization. More than five years after my studies on this topic began, I found the references I needed. A friend obtained for me copies of two books, out of print in their English editions, by the self-educated Russian scholar Konstantin Tsiolkowsky.[7,8] Born in 1857, Tsiolkowsky wrote pioneering works on reaction motors, multistage rockets and many other basic concepts of the space age.

Tsiolkowsky's novel *Beyond the Planet Earth*, written at the turn of the century, serialized, and finally published in book form in 1920, is a thinly veiled treatise on basic physics. As such it is short on characterization, and should be read for what it is: a daring but logical feat of the imagination. At a time when transportation was still almost exclusively horse drawn, it required a bold thinker indeed to speak casually (and accurately) of the necessary orbital speeds of kilometers per second.

As a novelist, Tsiolkowsky could skip lightly over the problems whose solution he could not then see — the rocket on which his voyagers lift off from Earth is powered by a mysterious explosive of a nature left unexplained. But the circumstances of the flight show surprising parallels to our present predicament on Earth. Tsiolkowsky postulates an Earth on which a growing population is beginning to feel the ecological limits. His travelers visit the Moon only incidentally; they realize from the start that the place for settlement is well away from any planetary surface:

> "Meanwhile the new colonies, five and a half Earth radii or 34,000 kilometers away, grew and were peopled. Mansion-conservatories of the type we have described were filling up with fortunate men, women and children. . . ."

They see the advantages of free space for establishing gravity convenient for particular tasks:

> ". . . nothing could be simpler than to create it artificially, you see, by rotating the house. In space, once you start a body rotating, it goes on rotating indefinitely, there is no effort involved; so the gravity is also maintained indefinitely, it costs nothing. Moreover, the amount of gravity depends on us; you can make it lower than terrestrial gravity, or higher."

On their first flight, Tsiolkowsky's travelers foresee accurately many of the possibilities of industry and habitation in space:

> "The space around the Earth which we can use — assuming we count only half the distance to the Moon — gets a thousand times more solar energy than the Earth . . . it only remains to fill it with dwellings, greenhouses — and people. By means of parabolic mirrors we can produce a temperature of up to 5,000 degrees centigrade, while the absence of gravity makes it possible to construct mirrors of virtually unlimited size, and consequently to obtain foci of any area we choose. The high temperature, the chemical and thermal energy of the Sun's rays, not weakened by the atmosphere, makes it possible to carry out all kinds of factory work, such as metal welding, recovering metals from ores, forging, casting, rolling, and so forth."

Sensibly enough, the travelers spend much of their first voyage in a search for usable asteroids. As a novelist, Tsiolkowsky has no difficulty in filling the asteroids with gold, platinum and diamonds, but in our more practical day we will be glad enough to find there such homely elements as carbon and hydrogen. Of all the prophecies Tsiolkowsky made during his long life, I am glad that one in particular was selected for the obelisk marking his grave in Kaluga:

> "Man will not always stay on Earth; the pursuit of light and space will lead him to penetrate the bounds of the atmosphere, timidly at first, but in the end to conquer the whole of solar space."

# *Chapter 5*
# *Islands In Space*

While we are considering the form of the new habitats for humans — the islands in space — we must always remember that details will change, perhaps profoundly, between earliest conception and final realization.[1,2] There may be better solutions to some technical problems than those I outline, and also new problems may arise whose solution will require changes in the design. I am describing a kind of "existence proof" — an illustration that a consistent solution does exist for the design of the islands in space. But it would be strange indeed if the efforts of one man were not improved upon greatly when others consider the problem.

I confess to a humanitarian bias in the design that I suggest. Technological revolution is a powerful force for social change, and in choosing among several technical possibilities I have been biased strongly toward those which seem to offer the greatest possibilities for enlarging human options, and for breaking through repressions which might otherwise be unbreakable. Yet I offer no Utopia. Man changes only on a time scale of millennia, and he has always within him the capacity for evil as well as for good. Material well-being and freedom of choice do not guarantee happiness, and for some people choice can be threatening, even frightening. Though I acknowledge that my study will be of the physical environment, and only indirectly with the psychological, I will still try to describe an environment which combines with its efficiencies and its practicality opportunities for increasing the options, the pleasures, and the freedoms of individual human beings.

I have argued that there is only one way in which we can develop truly high growth rate industry, able to continue the course of its development for a very long time without environmental damage — to combine unlimited solar power, the virtually unlimited resources of the Moon and the asteroid belt, and locations near Earth but not on a planetary surface.

I will describe first a community of what I like to call "moderate" size. It is larger than the first model habitat, but far below the dimensions of the largest that might be built. Island Three is efficient enough in the use of materials that it could be built in the early years of the next century. The numbers will seem staggering, but they are backed by calculation: within the limits of present technology, Island Three could have a diameter of four miles, a length of twenty miles, and a total land area of five hundred square miles, supporting a population of several million people. The largest communities

Small agricultural cylinders can ring the large main habitat. Within easy reach of the habitat, each can have a separately controlled environment and growing season.

that could be built, within the limits of ordinary, present-day structural materials like iron and aluminum, and with oxygen pressures equal to 5,000 feet above sea level on Earth, could be as much as four times as long and wide, with a land area half the size of Switzerland. It would be uneconomic at first to build habitats that large — they would be wasteful of materials. In the long run, though, the human race may build habitats of that size, or, with more advanced technologies, even larger.

The Island Three habitat provides three valleys, each 20 miles by 2 miles and opposite a large window called a "solar." Variable mirrors control the sunlight to each valley.

We need to provide gravity, water, land, air and natural sunshine in an Earth-like environment. Rotation can simulate gravity, and fortunately there are at least two geometries that allow rotation while giving us the real Sun stationary in the sky. One is a coupled pair of cylinders, whose long axes are parallel to each other. The cylinders are closed by hemispherical end-caps, and contain oxygen. Each cylinder rotates about its long axis, so that people living on its inner surface feel an Earth-normal gravity.

The cylinder circumference is divided into six regions, three "valleys" alternating with three arrays of windows. By locating three large, light planar mirrors above the windows, and pointing the cylinder axes always toward the Sun, we can arrange that the valleys will receive natural sunshine, and that the Sun will appear motionless in the sky even though the cylinder is rotating. Varying the mirror angle will give dawn, the slow passage of the Sun across the sky during the day, and sunset. The day length, weather, seasonal cycle and heat balance of the colony can be regulated by the same schedule of mirror angle variation. A large paraboloidal mirror at the end of each cylinder can be collecting solar energy twenty four hours per day, to run the community's power plant.

If we then set up many smaller cylinders near the big ones, and use the small ones for the growing of crops, we will achieve what has never been possible on Earth: independent control of the best climates for living, for agriculture and for industry all within a few miles of each other.

The "valley" areas in Island Three would each be two miles wide and

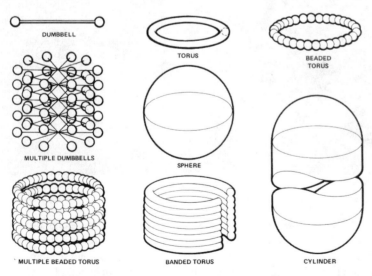

This comparison of the various habitat geometries shows the relative volume, surface area and Earth-normal gravity surface area available in each.

twenty miles long, rising beyond that to mountains. These mountains, formed on the inner surfaces of the cylinder end-caps, could have a height of up to 10,000 feet.

In the simplest version of a space community design, sunlight will be reflected into the habitat by large plane-surface mirrors, attached by many cables to each rotating cylinder and rotating with it. A dweller in one of the valley areas will look up and see a blue sky, obtained probably by art rather than by nature — it will be rather easy to control the reflectance of the mirrors and the tinting of the windows areas ("solars") to produce the most pleasing combination of warmth and brightness for the sunshine falling on the valleys, and to give a blue tint to the solars. There will be no sensation of rotation, though the cylinder will be turning once every two minutes. Gravity in the valley areas will be Earth-normal. No one in the space habitat will be in any doubt as to where he is, though — high above him, far above the clouds he will see, dimmed by distance, the other two valleys of his home. From that far away they will be as indistinct in detail as the Earth's surface is from an aircraft four miles high, but the inhabitants will be able to see them.

The angle of the sunlight entering the habitat will be controllable, and will depend only on the lengths of the cables which hold the mirrors. As the mirrors slowly open in the morning, the Sun will rise, but will move in the sky only as fast as it does on Earth. There will be no suggestion from its appearance that the cylinder is actually rotating. Only with very delicate instruments could one find that the image of the Sun's disc is rotating around its center.

With control over the angle of the Sun in the sky, the residents of space will also have control over the lengths of their days, the variation of the day length, and therefore the average climate and the seasons. They are unlikely to indulge in any sudden or capricious changes in those variables. Humans can adjust quickly, as the jet age has shown us, to changes in the day/night cycle and the climate. Plants and trees, though, are not so adaptable, and once a cycle has been established there will be good reason to make changes in it only very slowly.

By the time a community as large as Island Three is built, space habitats may not be occupied at the ecological limit — the highest population density that the land can support. In the early years of the next century Earth will be from two to three times as crowded as it is now, and the population density in space habitats may be falling toward the same value as that of Earth, ultimately to cross it and fall still lower. Island Three however could support quite easily a population of ten million people, growing its food in agricultural cylinders near but outside the main habitat. In the figure of habitat cost per person, I will assume that higher density. We are used to the perpetual conflict, here on Earth, between industry, agriculture and living space, but we

Agricultural areas, whether in many small cylinders or in a banded torus configuration will be highly automated, requiring minimal human labor.

must realize that in a space habitat economics will dictate escaping that conflict by locating agriculture a few miles away from the living areas. It is relatively expensive in materials to build large cylinders, with diameters of several miles, and relatively expensive to provide sunlight of normal appearance. Plants don't need such luxuries, and can be grown very efficiently in places where the solar intensity is high, but where there are no visual amenities.

With industry and agriculture located outside, the dwellers in Island Three can use their two hundred and fifty square miles of land area for living space and recreation. I suspect that as colonists from various countries of Earth arrive to settle the many communities in space, there will be a great variety in the ways in which land area will be used. Some immigrants may choose to arrange their land area in small villages, with single family homes, the villages being separated by forests. Others may prefer to build small, intimate towns of high population density, to enjoy for example the color and excitement and human interaction that is so much a feature of small villages in Italy. With many new communities to choose from, the emigrants from Earth will settle in those they like best. I would have a preference, I think, for one rather appealing arrangement: to leave the valleys free for small villages, forests and parks, to have lakes in the valley ends, at the foot of the mountains, and to have small cities rising into the foothills from the lake shores. Even at the high population density that might characterize an early habitat, that arrangement would seem rather pleasant: a house in a small village where life could be relaxed and children could be raised with room to play, and just five or ten miles away, a small city, with a population somewhat smaller than San Francisco's, to which one could go for theaters, museums, and concerts.

For Island Three, taken as a community that might be built in the first half of the next century, I assume that the population density will be "high," though as I have described it the habitat need not seem crowded. To feed that population, it will be quite enough to have an agricultural growing area equal to that of the living habitat. That may seem surprisingly small — it corresponds to growing the food to support one person on a plot only about thirty feet on a side. On Earth agriculture never reaches so high a productivity. The number is based, though, on yields which have already been achieved in a remarkable series of experiments by a unique individual.

After a long and active career at Cornell University, Dr. Richard Bradfield retired in 1965. Soon afterward he came out of retirement and assumed a responsibility and a physical challenge that many younger men would have found too demanding: the directorship, under the sponsorship of the Rockefeller Foundation, of the International Agricultural Experimental Station in the Philippines. That Institute, an outdoor laboratory for the development of new methods in high yield agriculture, is a world center for what has come to be called the "green revolution." Dr. Bradfield found that yields could be raised greatly by two expedients — multiple cropping and double planting.[3] In multiple cropping, advantage is taken of the fact that a high growing crop, like corn, can thrive in the same rows in which a low crop, like sweet potatoes, is grown. As long as nourishment is supplied to both by intense application of fertilizer, the two crops can live and grow together in harmony.

Double planting takes advantage of a fact that even amateur gardeners know: in the first weeks after seeds are planted, their growth does not depend on sunlight, or even on nutrients; all they require is warmth and moisture. The technique of double planting consists simply of overlapping one growing cycle with another. For a fast growing crop like hybrid corn, which reaches maturity in just ninety to one hundred days, the seeds of the next crop are planted ten or twenty days before the previous one is harvested.

By these methods Dr. Bradfield was able to reach very high yields even with what was basically conventional agriculture — not hydroponics. His agricultural station could support about twenty five people per acre, even in the less than ideal Philippine climate.[4]

Using Dr. Bradfield's data, we can work out the yield of agriculture products for a space community agricultural area, where the temperature will be ideal for growing (probably about like that of a hot summer day in Iowa) and the climate will be unvarying. Under such conditions there is no reason why four crops per year cannot be obtained.

In the developed countries we are used to a varied and probably too rich diet, but in the space habitats there is no reason why anyone should be limited to a diet of rice or cereals. Specialists in diet tell us how many calories and how many grams of usable protein we need every day if we're doing active physical work, and the space agriculture areas are figured on that basis.[5] Many of us find that on such a diet we have trouble keeping from gaining weight. In the early space communities it will not be very practical to raise many beef cattle; they are quite inefficient at converting plant foods to protein-rich meat, losing a large factor in the process. Chicken and turkeys, though, are quite efficient, and pigs are not much less so. Dr. Bradfield has found that in high yield agriculture the cuttings from crops like corn and sweet potatoes can be used efficiently as pig fodder,[6] and that practice can be employed to good effect in a space community's agricultural cylinders.

With a varied diet including all the corn, cereals, breads, and pastries that many of us enjoy, and with plenty of poultry and pork, the space colonists will have good reason to follow our Pilgrim ancestors, and celebrate Thanksgiving with a feast of turkey, and Christmas with a savory ham. There will be no need for anyone to think in terms of pressed soybean cakes or fishmeal unless they happen to like such things. With four crops per year, a completely dependable climate free of hurricanes and frosts, and the techniques that Dr. Bradfield has developed, the space communities can easily support at least fifty three people per acre of agricultural land.

The agricultural areas of a space habitat will probably be relatively small, perhaps a square mile in area. They may be cylinders, but will not have external rotating mirrors — simple conical reflectors will be quite sufficient for them, since a stalk of corn will hardly care whether the image of the Sun is round or elliptical. The agricultural areas will probably be run at rather a low density of oxygen, corresponding perhaps to a mountain altitude, because that will make the enclosing structure cheaper and because plants grow best with less air. The climate will be hot and moist for most crops, and the day length will be controlled in an inexpensive way, by drawing an aluminum foil shade in zero gravity, outside the cylinder, across the front of the mirrors to close off the sunlight. Looking into such an agricultural cylinder, one would rarely see a farmer, just as it is rare to see a human being as one drives through the San Joaquin Valley, one of our highest yield agricultural areas in the United States. As in that valley, the occasional farmer will almost certainly be driving a machine of some kind, a planter or harvester. At the space community the machine may be air conditioned, perhaps even pressurized, and shielded against the radiation from solar flares.

When winter comes to the San Joaquin, it closes down the growing season over the entire valley at once. That will not happen in the space habitat. There, each cylinder can have the climate and season that people choose, because those factors will be controlled by the moisture content of the air and ground, and by the schedule of day length for each cylinder. With such control, there is every reason not only to tailor the climate of each cylinder to favor its particular crop, but to distribute the seasons among the cylinders. They may even run in serial order: January, February, March, and so on. With that degree of freedom it will be

In each of the three long valleys in Island Three the "sun" appears round and overhead, just as on Earth, as the positioning shrouds on the mirrors control the "day."

possible to have every desired crop always "in season" in one of the areas, so that the settlers living a few miles away can enjoy, for example, fresh strawberries even in the middle of what may be their January.

In the long run, when plentiful supplies of water are available from the asteroids, it will be possible to specialize certain of the agricultural areas as ponds and lakes, both fresh and salt. Oysters, clams, fish of all kinds, and perhaps even that vanishing delicacy the lobster, may all be grown there.

The practicality of these options depends on the free solar energy continuously available in space, and on the fact that this energy can be used for the inexpensive production of chemical fertilizer. A plant using direct thermal energy and converting oxygen and nitrogen to nitric oxide, adequate to provide abundant fertilizer for a space habitat, need have a concentrating mirror area of only a square meter per person, because of the high intensity of sunlight in space and its availability day and night, all year.

I like to be on the safe side in my estimates, and by the time space colonies become a reality it's likely that we'll be able to do even better than I've "promised." Detailed studies supported by NASA in 1975 and 1977, with participation by experts in high yield agriculture, concluded that the numbers I've given are well on the conservative side. The General Electric Company feels sure enough about closed environment greenhouse agriculture that in 1977 it committed corporate funds for a half acre pilot plant, and is betting on yields even much higher than I've figured.

The 20 mile long valleys in Island Three will have clouds overhead and will view the stars, Earth and the Moon at "night." Even an occasional solar eclipse is possible.

The space habitats will operate, of course, in a totally recycling way — fresh produce, fruit, vegetables, meat, milk and cheese will travel the short distance from the agricultural areas to the living habitat, and the return flow will be pure water and nutrients for the fertilizer plants — nothing will be thrown away. Passing all wastes through a high temperature solar furnace, which will cost almost nothing, will ensure that everything entering the agricultural areas is sterile. In that way they can be kept free of pests, even if any should accidentally be introduced into the living habitat. In the very worst case — the introduction or evolution of an agricultural disease in one of the areas — the sterilization process that will be part of the recycling will ensure that the disease will not spread. As soon as such a disease is found, there will be a simple and preferable alternative to the Earthbound necessity of sprays and poisons: it will only be necessary to drain off the water of the contaminated cylinder through a solar steam boiler to a sterile tank, and to open the shades so that the cylinder heats up to a temperature at which no living organism can survive. After a few days or weeks of that treatment the water can be re-introduced, the appropriate soil bacteria can be replaced, and a new planting cycle can begin.

The population density in the space habitats will be governed by sheer economics. There will be a certain cost per square mile of land area, low for the minimal, early communities, higher for the larger ones. A habitat of large diameter will require a thicker supporting shell of aluminum or steel. As I have emphasized, a key element in the humanization of space will be the unchecked continuation of the industrial revolution, the process by which average individual productivity and wealth increases. That increase translates into time, when we consider population density in a community — in the early stages it will not be possible economically to build and amortize a community unless it contains a large work force to pay off the construction costs within the amortization period. Later, as automation, productivity, and with them average wealth increase, it will become possible to build relatively large communities for habitation by comparatively few people. As we shall see in a later chapter, that transition will not take long. With a normal growth in productivity, less than a century will suffice for a reduction of a factor of ten in population density.

Island Three, though, is taken to be an early model, built when productivity is still not very much greater than in a developed nation on Earth at the present time. It may have ten million people, and we should look at what that large population will mean in terms of living conditions.

With half the total population living in small cities on the mountain slopes, and agriculture carried out in the two hundred seventy square miles of external cylinders, the habitat valleys may be used entirely for green areas and for suburban towns. Though the life styles may be as varied as the national origins of the colonists, one possibility is to have a series of small villages nested in a forest. A population of 25,000 in a village will be enough to support schools and shops, and such a town need be little more than a mile across. With predictably good weather and a mild seasonal variation, bicycles and small electric runabouts will be quite adequate for travel within the village, so it can be a place free of automobiles and of the internal combustion engine. Though I speak of Island Three as a high density community, the living conditions in the village will hardly seem crowded. A family of five could easily have a one story home of four or five bedroom size, with large living areas. They could have, in addition, a garden and a yard of equal area, while still leaving much of the village free for shops, schools, and perhaps a village green.

Some features of the habitat geometry will lead to new possibilities in house design. For one, the ubiquitous, ugly TV antenna of American suburbia will vanish, to be replaced by a small built-in concealed equivalent, pointing at the center of the cylinder endcap. With direct line-of-sight and a distance of only a few miles, reception should be superb. Probably by the time such a community is built families will also be able to communicate with a central library through the same microwave link.

Electric power, brought underground from the external solar power station over cables laid in when the community is built, will run lights, appliances and air conditioning. Most of the energy we use, though, goes into heat, for house heating and for cooking. In Island Three, all such directly used heat may be obtained from solar power, without ever going through an intermediate stage as electricity. The ground on which houses are built may be no more than two feet thick, and at the time of construction several access channels down to the outer shell may also be built in. Even at nighttime solar power will always be available, no more than a few feet from the house floor. Reflected by external mirrors, solar heat for cooking can be brought up through the floor through a window and through a short channel, to be absorbed on the lower side of a simple metal cooking surface. A rather powerful electric range element can be replaced by a cooking surface fed by just two square yards of solar collecting mirror, and turned off by a simple shutter. The heating of each room in the home can be done in the same way. Whether that is the course adopted, or whether electric power will be so cheap that it will be used for all energy needs, is a matter for economics and design ingenuity to determine.

The homes of Island Three may also have a design detail no Earthbound home can ever equal: a window set at an angle in one wall of a living room, through which the immensity of space and the brilliant, unclouded stars will always be visible, drifting majestically across the field of view as Island Three rotates in its unvarying two minute cycle.

The production facilities of Island Three may be of two kinds: light industry, located in the cities or even in the villages; and heavy industry, outside the habitat entirely. On Earth, industry must compete with us for land area on which to locate. But no such conflict will arise at Island Three.

An industrial complex located just outside an end of the community, and nonrotating, will be an ideal facility for processing lunar materials into finished products. At each end of each cylindrical habitat, there can be a thin, nonrotating disc of zero gravity industries — a disc as big as the colony's end-cap, but only as thick as one factory. In that arrangement, each industry can have its own direct access to space, to receive raw materials shipments and to send off its finished products. In such a geometry waste heat from these industries can be radiated away into the cold of outer space with equal ease. Workers in these zero gravity industries can travel from the cylinder axis to their jobs in a few minutes, in a large air filled zero gravity corridor, pushing off from their starting points and drifting in free flight to their destinations — and possibly reading their magazines as they go.

The products of these zero gravity industries could be very large indeed. There is no reason why an external factory couldn't build and fully assemble a complete solar power station, which could then be gently floated away in zero gravity to its destined point of use.

In the energy rich environment of a space community, it will normally be more efficient to recover and separate industrial waste products for their useful materials, but if any smoke or gases do escape from a factory they will be carried by the solar wind all the way out of our solar system, never to add pollution to the environment.

In most of the agricultural areas of Island Three, except for the insects essential to pollination, there will be no reason to have animal life; no birds, for example, because they would attack the crops.

In the main living areas, though, we may find the ideal habitat in which species endangered on Earth may survive. There will be no need to introduce insecticides or other poisons, and industrial wastes, if any, will be borne away by the solar wind, never to enter the habitat itself. In those conditions, with choice in the species which are introduced to form the initial ecosystem, it may be quite possible to bring rare species of birds and animals from Earth to the nonagricultural areas, and to have them survive and flourish.

Every step toward the settlement of space will benefit conservation programs in another way — by relieving Earth of industry and of its burden of population, so that the species of animals, birds and fish now in danger on Earth will have a better chance for survival here.

The valleys of Island Three are primarily residential areas, with perhaps some light industry. Landscaped grounds and park area will be in abundance.

# *Chapter 6*
# *New Earth*

A few hours of time with pencil and paper, while letting the imagination roam, will be enough to convince any reader that many geometries are possible for habitats in space. In the long run, it seems likely that in designing their environments dwellers in space will take full advantage of new degrees of freedom in gravity, day length, and climate. My reason for describing a much more conservative, narrowly Earth-like habitat is that all of us now on Earth, who must decide our priorities for the years ahead, must do so on the basis of well known ways of living, methods that now work for at least a part of the human race. Our descendants, raised from birth with zero gravity and adjustable seasons as commonplace elements in their lives, will be far more inventive than we are in turning those options to advantage. Those who inhabit the first communities will not have that head start. They'll find enough "future shock" as it is, in making the transition from Earth to a space habitat, and for them it may be reassuring to know that they may look forward to something familiar and homelike.

In that vein, it's interesting to consider some of the possibilities for modeling directly attractive portions of the Earth. We think of a valley area two miles by twenty as rather small, but it is surprisingly large when compared to some of humankind's favorite places. Most of the island of Bermuda, including the lovely south coast areas named after the English shires, could be modeled rather well within about half the length of a space community valley. Only when population densities have dropped and a plentiful water source is available from the asteroids will that sort of whimsical luxury be possible. There's a small but lovely bit of the California coast, including the town of Carmel, favored by artists, writers, and many visitors. The usable area of an Island Three space community would be more than twenty five times larger. We may expect that, in common with our ancestors who chose wistfully to call their frontier "New England," at least some of the settlers in space will model their cities and villages on the prettier areas of Old Earth.

As little as a year ago, I would have felt it necessary to write at considerable length on the structure of the habitats — the aluminum or steel cables, or the metal shells which band them and contain the forces of atmosphere and rotation; the solars which admit the sunlight while retaining the atmosphere. There is no need to do that now. A number of engineers in several government and private industrial laboratories have checked the relevant calculations. It's enough to say that the construction techniques are not basically new, being for the most part variations on the methods of Earthbound bridge building or shipbuilding. The strength assumed for aluminum corresponds to well known alloys, with the safety factors that come out of the standard engineering handbooks. For steel cables the numbers are within normal practice for suspension bridges, higher but not double the value that was common in terrestrial bridge building as much as fifty years ago.

One problem in basic physics is worth discussing, though, because one of its possible solutions brings with it a number of diverting possibilities which the space dwellers may want to exploit. A rotating cylinder in space constitutes a gyroscope, and in the case of a space community it is an enormous gyroscope indeed. As we learned in school, a gyroscope left free will continue to point its rotation axis always in the same direction, relative to the distant stars. That is the principle on which the gyrocompass works. In the case of a space habitat, gyroscopic action could be a problem. The simple use of solar power, as well as the arrangements for natural sunlight and for the day / night cycle, all require that sunshine always arrive in a direction along the cylinder axis. One way to satisfy that condition is to orient the cylinder axis perpendicular to the community's orbit around the sun, and to provide a lightweight mirror, angled at forty five degrees, to bring to sunshine along that axis.

Alternatively, the cylinder axis can be in the plane of the orbit. In a year, as the community moves with the Earth around the Sun, in that case the axis must turn through one complete rotation. In order to provide that turning motion of the cylinder axis — the motion physicists call precession — forces must be applied. The forces need not be large, because the precession rate will be very slow: only about one degree per day. Calculation shows in fact that the forces will be only about one ten millionth of the weight that the cylinder would have at Earth's surface. Two equal and opposite forces are needed. They can be applied through hollow bearings at the cylinder ends. The bearing forces will be small in comparison to the shock loads on a locomotive wheel or on an aircraft while landing, so the bearings themselves will not be difficult to build and need not be large. At each end, the forces can be taken by tension or compression towers, as thin and spidery in appearance as terrestrial radio masts.

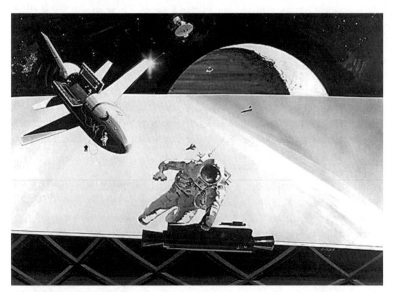

Construction techniques in space are not basically new, being for the most part variations on the methods of Earthbound bridge building or shipbuilding.

Where now can we obtain the lever to move this small world? One easy way is to obtain the required forces by attaching the towers to another cylinder, as nearly as possible identical in mass and size to the first. In that way each can supply the necessary forces for the other. One habitat can be located above the plane in which Earth moves around the Sun, and the other just below it. We can convince ourselves of the correctness of this solution by considering that when the two cylinders rotate in opposite directions, their net gyroscopic action will be zero. As such, there will be no resistance to turning them as a pair, and once they are established in that slow precession they will hold it to eternity without the need for more than small corrections.

For reasons of necessity, to satisfy the equations of mechanics, it seems the space dwellers may adopt a geometry that links two habitat cylinders together a form a complete community. No energy or power will be required for that arrangement, and no rocket thrusts will be necessary to obtain the necessary forces, so the solution should be inexpensive. The tension member, in fact, need be no bigger in diameter than a teacup.

If they adopt this solution, the colonists will now find that it provides them with some free benefits. The first concerns seasonal phase: the mirror schedules of the two sister habitats will be independent, so the seasons in the two may be as different as the inhabitants wish. One may be in the midst of January weather while the other is in June. Another possibility will be to have a rather severe climate for one of the two habitats, with seasonal extremes perhaps as great as those in New England; warm summers and clear, snowy winters, for skiing and for a Dickensian "White Christmas." The other habitat, only fifty miles away, may have a climate as lush and tropical as that of Hawaii. If travel between the habitats can be arranged to be easy and inexpensive, there seem to be attractive new options for visits crossing seasons or climatic zones.

Given the two rotating cylinders in space, parallel to each other and only fifty miles apart, the space dwellers will be able to take advantage of the rotation which produces Earth-normal gravity in the habitat valleys. For Island Three, that rotation is at a rate of about four hundred miles per hour. Imagine then a simple vehicle, less complicated even than a terrestrial bus: it contains comfortably spacious seating, but requires no engine or crew. As its passengers board it, walking down stairs through their land valley as if they were entering a subway station, the vehicle will still be locked to the outer surface of the habitat. When the door is closed and sealed, a computer on the habitat will wait until the correct moment in the cylinder's rotation cycle, then will unlock the vehicle. Proceeding through space on a straight line, with the tangential velocity of the habitat, the vehicle will arrive at the other cylinder in less than eight minutes. On release it will have been given a gentle twist, so that it will perform a half roll in the few minutes of its flight. On arrival, it will find the outer surface of the second cylinder moving at exactly its own speed, and will lock onto the outer surface in a dock similar to the one it just left. The passengers, after their few minutes of zero gravity, will feel weight restored, and can leave their seats, take the "up" escalator, and find themselves in what is literally another world, perhaps as different from the one they left as Polynesia is from Maine in winter. Such transportation should be quite inexpensive, because it will require no energy. That sounds vaguely disquieting, rather too much like perpetual motion, but in fact the statement is true — the transfer from one cylinder to the other, in such a vehicle, will require no motive power.[1] Given that convenient fact, and the high utilization obtained from a vehicle that can make several trips per hour, the cost of such a journey surely will be quite low. We can imagine young people "jet-setting" over from one habitat to the other just for the afternoon, complete with their snow skis or their water skis, for no more than the cost of a bus token.

Many of Island Three's inhabitants will be commuters, going to their jobs in the cities or in the zero gravity industries, from homes in the valley areas. With the new degrees of freedom that will exist in the space habitats, it will be possible for them to do their commuting far more comfortably and quickly than do Earth's tired millions of workday travelers.

The valleys form natural lines of communication joining the cities and their suburban areas. No village will be more than a mile from a valley center. For that short distance, bicycles or small electric powered runabouts of bicycle speed will be quite adequate. At the valley centers, it will be natural to have rapid transit systems, but there again a new option will be possible.

Within the past ten years several nations have begun investigating what is called "dynamic magnetic levitation." That is a lifting force on a vehicle which occurs when the vehicle is equipped with permanent magnets and "flies" above a conducting guideway.[2] The technology of high field superconductors, which has been brought to commercial practicality only within the last ten years, now makes it possible for a vehicle to sustain a strong constant magnetic field without the expenditure of power. If a vehicle is standing still above a piece of aluminum, it will simply fall when released. But if it is given a forward motion, the eddy currents which its field induces in the guideway will generate counterfields, which will act always to produce lift. Dynamic magnetic levitation has several advantages as a substitute for wheels and rails in transportation systems: it is efficient; it gives a soft, gentle ride even at high speed; and above all it does not require high precision in the location and leveling of its "track." The Magnetic levitation system, sometimes called "Mag-lev" or the "Magneplane," is inherently capable of high speeds, from two hundred to three hundred miles per hour. On Earth there is some difficulty about attaining those speeds, because of aerodynamic drag and the high noise level produced by a train cutting through the atmosphere at sea level at so high a velocity.

In a space habitat, magnetic levitation may come into its own, because high vacuum is an ideal dragless, noiseless medium through which a magneplane can travel at high speed. Probably the residents of space, arriving at a station within a mile of their homes, will set their electric vehicles to find their way home along the bicycle paths at a safe walking speed, following the magnetic lure of a buried wire. Entering the magneplane station, the commuters will go down through the habitat shell, and will board the vehicle when it arrives. The airliner doors will close, a diaphragm will close to seal off the entrance, and the magneplane will begin to accelerate in the high vacuum only a few feet below the valley, reaching in less than a minute a speed of three hundred miles per hour in silence. Minutes later it can decelerate for its stop at the city; or if scheduled for the zero gravity station near the endcap's hollow bearing, it will coast on its magnetic lift up the outside of the cylinder end hemisphere, stopping at a point where travel can continue without vehicles, in the drifting flight of zero gravity. With unlimited low cost electrical energy from the habitat power station, and with computer control over their movements, probably these efficient vehicles will be able to operate at intervals of only a few minutes, so that people wishing to travel to or from the cities or to the industries will not have to worry about timetables and can go whenever their own schedules make it necessary or convenient.

In an earlier chapter a method was suggested for the transport of the work force from a community to an industrial complex, which might be located at a distance of a hundred miles away. The reasons for such a location might be the need to isolate the habitat from the waste heat radiated from an intense user of energy, or the availability to an industry of workers who might choose, through personal preference, to live in habitats of different climate or architecture than those of the nearest community.

Much the same method should be usable for long distance transportation. A sphere much like the "commutersphere" I spoke of earlier, but with perhaps only half as many passengers, could provide luxurious conditions of comfort and convenience. While being accelerated the sphere could be given a rotation, leaving it with a fractional gravity to simplify such practical functions as eating or going to the toilet. I like Arthur Clarke's comment on the alternative, rapid acceleration and deceleration with zero gravity in between: "Half the time the toilet's out of reach — the other half it's out of order."[3]

As we've learned on Earth, speeds in the jet aircraft range are quite adequate for travel over intercontinental distances. On Earth, unfortunately, the conditions under which that travel goes on are cramped and uncomfortable. Aerodynamic factors and the need for an on-board crew — to cope with weather, mechanical failures and the tricky operation of landing — force the design of commercial aircraft that are too large to be intimate and too crowded to be restful.

For a flight of the same distance as New York to Los Angeles, but in an electrically accelerated "travelsphere" plying a course between space colonies, getting up to speed will take only a minute or so. The rest of the flight will cost nothing except that portion of the vehicle's initial cost and maintenance that are being amortized over that time — and, of course, the costs of food and cabin service. One very big difference between flight in the atmosphere and flight in space is that in space we don't have to worry about the speed of sound. The travelsphere can fly at higher than Concorde speeds, but with no concern about either sonic booms or pollution of the atmosphere. Making generous assumptions about the cost of the simple vehicle involved, and assuming load factors, utilization and amortization schedules similar to those of terrestrial jetliners, it appears that flight in a travelsphere might cost about a fifth as much per passenger-mile as the ticket cost on a modern jet like the Lockheed L-1011. It's strange to think that the travelsphere would be a much simpler vehicle, but those are the facts — no engines, no complex electronics, no complicated structure to take atmospheric buffetings. And, of course, no burn-up of scarce petroleum resources.

It's diverting to think what this sort of efficiency will mean for travel over shorter distances, also. In a commutersphere, for less than the round-trip cost of a short automobile trip on Earth, a man could take his wife to dinner in another community — a few minutes in the run-about, five minutes on the magneplane, a half hour flight in a travel sphere, and the couple could be in another habitat, after choosing among dozens of them located within that travel time of their own. That could mean selecting a concert or opera performance, or simply a dinner in a favorite restaurant, in a community which could be as different in culture and language as Rome is from Kansas City.

For good health we should spend some of our time in Earth-normal gravity. Yet much of the recreation in which the residents of space indulge will surely take advantage of a new option we can never experience on Earth: to have any gravity they like, simply by riding or walking to the right distance from the cylinder axis. On the axis itself gravity will be zero, and it will increase smoothly toward Earth-normal as the valley floor is approached.

Surely new sports will be invented to make use of this degree of freedom — three dimensional soccer may be one example. Some old sports will also be a great deal more enjoyable in low gravity. In a pool near the cylinder axis, a dive will be made in slow motion and the waves will break as slowly as in a dream. Those of us who enjoy scuba diving find that under Earth's oceans the need for pressure equalization reminds us, with every foot of depth change, that we are not in our natural element. A pool near the cylinder axis, or an entire sea world, perhaps in one of the external cylinders, could have a gravity as small as a thousandth that of Earth, and could give the swimmers of the habitat the freedom to forget pressure changes and swim as naturally and freely as the fish.

It seems unlikely that any of the communities will be willing to put up with powered aircraft, because of their noise and smoke, but soaring — the use of air currents to sail in three dimensions with a glider — should be possible. As a glider pilot I find that people even on the ground seem to feel a sensation of joy and release in watching a glider fly. As Richard Bach has said, perhaps there is something of Jonathan Seagull in each of us.[4]

From the time of classical Greece, and perhaps even before, some men have been fascinated by the idea of flight by human power alone. Leonardo da Vinci was obsessed by it, and filled notebooks with sketches of machines which he hoped might fly. In modern times man-powered aircraft have been flown short distances, but under Earth-normal gravity human powered flight remains an almost impossible dream. In space communities, it will become easy for everyone, not just for athletes. Close to the cylinder axes, in near zero gravity, almost every imaginable variety of human powered flying machine, including some of Leonardo's, will work. We can imagine elderly ladies and gentlemen taking their evening constitutionals by gently pedaling their aircraft, while viewing the world miles below them. Because they will be in a "gravity" produced by rotation, they will be able to change it at will, by flying with or against the direction the habitat is turning in. While as far from the axis as the height of a tall building, they'll be able to cancel gravity entirely by pedaling at only bicycle speed — but in the right direction.

As at swimming beaches, space dwellers may have to provide something to keep people out of danger. There are at least two possibilities: one is a near invisible cylindrical net to prevent a tired flyer from straying too far from the cylinder axis into a high gravity region. Another is a parachute, permanently mounted on the pedal plane, and ready to pop open if the flyer descends too low.

Where the valleys end and the hemispherical endcap begins its upward curve toward the cylinder axis, the temptation will be great to model the mountains of Old Earth. A hike up those mountains will be a good deal easier than on Earth. As the climber makes his way to higher altitude, and starts to become tired, gravity will be lessening with every foot of height gained. By the time he's two thirds of the way up the mountain he'll weigh only a third as much as he did on Earth or at the start of his walk, and can climb in bounding strides. At the top, two miles above the valley, he will weigh nothing at all. He will have passed the clouds at about the 3,000 foot level, so they will be far below him, but he will find that the atmosphere has lowered in density only as much as for a climb to half his elevation in the mountains of Earth.

I've devoted a good deal of this chapter to the less serious side of life in a space colony — not questions of economics and production, but of amusement and diversion. It seems appropriate to close with an account of one memorable lunch time conversation. In the years before the topic of this book was well known, I had made a practice of challenging skeptics to name their favorite sports, and then always pointing out that the sport could be done better in space than on Earth. Finally someone named a delightful sport that, even in these uninhibited days, is carried on only in private. The skeptic instantly became a believer — can one imagine a better location for a honeymoon hotel than the zero gravity region of a space community?

# Chapter 7
# Risks And Dangers

Almost every human activity carries with it some element of risk. Occasionally, in a rare macabre frame of mind, I have reflected on the fact that at any time almost every human being, however healthy, is within one or two minutes of death if the wrong combination of circumstances were to come to pass. When I lecture on the topic of habitats in space, it is natural that some of the questions that follow relate to the possibility of violent catastrophe in a space community. Given the fragility of life, that possibility will always be there, so we must be quantitative and estimate the risks that will attend the human settlement of space. It's reassuring to find that in fact they are rather less than those to which we are exposed every day here on Earth.

Almost invariably the first question that is asked about space habitats concerns meteoroids. These are, for the most part, grains of dust which have been in the solar system since its formation several billion years ago. As our Earth revolves around the Sun each year we travel at a near constant speed of about thirty kilometers per second — higher than any of the relative speeds needed for launching a satellite or traveling to L5, or even for voyaging to an asteroid. Most of the grains of dust which we encounter in our annual passage around the Sun are moving relatively slowly, so typical relative speeds with which we meet them are just our own. Almost the highest speed meteoroid which has ever been measured corresponds to a dust grain moving in a circular orbit around the Sun, but in a direction contrary to our own; combined with our own velocity that gives an encounter at doubled speed.

Most of these meteoroids are of cometary rather than asteroidal origin, and can be thought of as dust conglomerates, possibly bound by frozen gases.[1] If present scientific ideas are correct, therefore, a typical meteoroid is more like a mini-snowball than like a rock. Even a very small meteoroid carries, because of its velocity, a great deal of energy, but fortunately almost all meteoroids are of microscopic size. In the frequency curve of their occurrence, as the size increases their number goes down rapidly. Spacecraft sensors have collected abundant and consistent data on meteoroids in the range from one gram (that is about one thirtieth of an ounce) down to a millionth of a gram.[2] Above that size, there is so small a chance of finding a meteoroid that even in a voyage of years a spacecraft records almost no data.

For relatively large meteoroids, the series of Apollo flights has left us with a scientific legacy especially important for just this question — the Apollo seismic network, a series of very delicate seismometers left on the Moon. These instruments continued to record for many months after the flights which installed them, and they have recorded not only Moonquakes but the collisions of meteoroids with the lunar surface. So sensitive are these machines that their builders claim to be able to detect every strike occurring anywhere on the Moon by a meteoroid of soccer ball size or larger. Fortunately these two independent means for measurement of the meteoroid size distribution agree quite well, and allow us to estimate with some accuracy the chance of a strike on a space habitat, for a meteoroid of any given size.

There is a third method for the measurement of meteoroid size distribution. It is ingenious and relatively inexpensive: an array of wide angle cameras, forming a pattern which is called the "Prairie Network" is distributed over about one million square miles of lightly populated farming states in the central part of the United States. When a meteoroid enters our atmosphere, leaving the luminous trail which we call a meteor, the Prairie Network sky cameras photograph the trail with such accuracy in space and time that the position, altitude and velocity of the meteor can then be calculated. Some of the best measurements of speed distributions come from data of this kind.[3] Unfortunately, it is much harder to obtain from that source accurate figures on size distributions. Those have to be based on the brightness of the trails observed, and then on a crucial assumption: how much of the energy of the incoming meteoroid is converted to heat and light.

The Prairie Network data agree with those of the other two methods quite well for meteoroids the size of a marble. They aren't in such good agreement for the larger or smaller ones, probably because of the assumptions made about luminous efficiency. If one assumes, as is consistent with the most common modern view, that the typical meteoroid is a dust conglomerate, then the efficiency of conversion of the incoming energy to heat and light should be rather high. With that assumption the camera data agree better with those of the other two methods than they do if a low efficiency is assumed.

Averaging the data from what seem to be the most reliable sources, one finds that in order to be struck by a meteoroid of really large size, one ton, a large Island Three community would have to wait about a million years. Such a strike should by no means destroy a well designed habitat, but it would certainly produce a hole and cause local damage.

In order to find meteoroids that would strike at a frequency high enough to worry about, we have to consider much smaller sizes, of about the weight of a tennis ball. On one of the big communities, there'd be a strike by one of those about every three years. Curiously, there is a reason why a habitat of given size would be struck less often than an equal area at the top of Earth's atmosphere — the gravitation of Earth is so strong that it "sweeps out" meteoroids, sucking them in from a region of space much larger than its own area. The space habitats, far enough away from Earth not to be in the affected region, and having almost no gravity of their own, would be stuck relatively less often.

The most vulnerable parts of a habitat will be its windows. They will occupy a large area and, being made of glass, will be relatively fragile. They will naturally be subdivided into small panels, for two reasons: to guard against the possibility of catastrophic damage, and to allow the aluminum, steel, or titanium supporting structure to carry all the structural strength in the window regions. A window panel may have an area two or three times that of a window on a jet aircraft. With such a size, the metal frames that carry all the structural loads can be so thin that they will be invisible from a valley floor, and the windows will appear continuous when viewed from that distance.

For panels of that size, the loss of one will certainly not be catastrophic for the community. For what we have called Island Three, if one panel were blown out entirely it would be several years before the atmosphere would leak out. Detection of a blowout should be almost instantaneous — it would result in a plume of white water vapor, condensing to ice crystals in vacuum, visible from the sister habitat. If a patch were put on the blown out panel within an hour, the loss of water vapor would be economically tolerable (the oxygen would cost far less to replace) and probably no one but the repair crew would even know of the event.

Even for the smallest community, Island One, the corresponding numbers would be quite tolerable. There it would be several thousand years between strikes by a meteoroid big enough to break a window panel. When a panel blew, if it were patched within an hour the loss of atmosphere would reduce the pressure by only about as much as we would find on Earth in climbing a hill two hundred feet high — not even enough for us to detect a pressure change on our eardrums. For the most recent design of Island One, these risks would be further reduced by a large factor. We now assume a design in which heavy shielding, provided for cosmic ray protection, would protect the window areas from any direct "view" of space.

At the surface of the Earth we are exposed to radiation from three different sources: emanations from the soil, rocks, bricks, and other structures which make up our environment; radiation from small quantities of radioactive substances within our own bodies; and cosmic rays which penetrate our atmosphere. Radiation is measured in units of Roentgens, and for biological damage the unit rem (roentgen equivalent man) takes account of the differing amounts of damage done by radiations of various kinds. For total dosage over a period of time, the unit is the rad (radiation dose). On Earth's surface the amount of radiation to which people are exposed varies over an enormous range, depending on where they live.

Oddly enough, most of the radiation the average person gets comes from inside — trace amounts of radioactive elements in the body. The radiation from outside depends on such details as whether one lives in a brick house (bad) or a wooden house (good). Most of all, though, it depends on geographical area. In the monazite sands region of India the residents get a natural dose of almost one rad per year.[4]

By comparison, our normal dose from cosmic rays is relatively small — least of all at sea level near the equator, but still only a small fraction of a rad per year for a mountain elevation in a temperate latitude. At the poles it is much higher; the latitude differences arise from the fact that Earth possesses a magnetic field which provides it with a substantial amount of protection against the lower energy cosmic rays.

When all the sources of natural radiation, internal, external and cosmic, are added, they amount to an average dose of about a third of a rad per year for a typical Earth dweller. After a great deal of testing and years of discussion, to which many physicists and biologists contributed, the Atomic Energy Commission (in the days long before it was called ERDA) settled on an allowable annual dose for its workers of five rad per year, and of a tenth that for the total U.S. population.

Clinically, only the most sensitive and delicate laboratory tests can detect effects in humans from average radiation of less than about twenty rad per year, and far larger average exposures are required before a human individual is aware of any consequent illness or discomfort.

In space, far from the protective shield of Earth's magnetic field, the level of steady, highly penetrating cosmic rays (the so-called primary galactic radiation) is about ten rad per year. If there were no other radiation to consider, it would be reasonable to consider building the first space habitats with no shielding at all.

If a large fraction of the world population were to live in those conditions for many centuries, we should be concerned about the resulting increase not only in cancer but in the rate of mutations. That would not occur, though — the buildup in the size of habitats to the point of thorough shielding would take place over at most a few decades of time, and during that brief time only a small segment of the human population would be exposed to enhanced radiation levels.

There is however a more serious cosmic ray problem, arising from a type of radiation to which we are never exposed on Earth. These rays are the "heavy primaries": nuclei of helium, carbon, iron and the whole range of elements found on Earth. They form only a tiny fraction of the total cosmic radiation, but they are far more damaging than the rest.

When heavy primary cosmic rays pass through material, they leave a dense trail of ionized atoms. These atoms are highly active chemically, and are so numerous that in living cells they cause cell death. The same property of intense ionizing power which is responsible for the biological damage done by heavy primaries is also a protection against them — in our atmosphere they lose energy so quickly by ionization that they are absorbed at high altitudes, never penetrating to sea level.

The only direct human experience with heavy primaries has been that of the Apollo astronauts, who ventured outside not only the atmosphere but also the protective magnetic shield of Earth. In that open region they observed flashes of light, visible especially when they adapted their eyes to total darkness. Most scientists who have studied the subject agree that these light flashes were almost certainly caused by heavy primaries. On Apollo 17 a systematic study was made of this effect. When I asked Dr. Harrison (Jack) Schmitt, who went to the Moon as an Apollo 17 scientist-astronaut (and later was elected U.S. Senator from New Mexico) about his observations, he reported an odd fact: although the light flashes were visible at a rate of one every few minutes throughout most of the voyage, during the period of one deliberate experiment none were seen for an interval of an hour or so. At present no one has come up with a good explanation for how they could have vanished, even temporarily.

On Apollo 12 the astronauts were exposed to the heavy primaries for about two weeks. Estimates based on direct radiation measurements and the known sizes of body cells suggest that during that period their loss of brain cells was a few in a million. A similar figure holds for retinal cells, and for the very largest body cells (neurons) the fraction is perhaps as much as one in ten thousand.[5] These are small numbers, but there is still reason for concern about them — the cells involved are nerve cells, and as such are not replaced by the normal body repair mechanisms. We have then one "data point" which we could take as conservative for our further calculations: the Apollo 12 crew was exposed to a certain known dose of the heavy primaries, and suffered no apparent ill effects from them. To be on the safe side, therefore, our design of even the first space habitat should be based on the requirement that in a working career of several decades a human being would be exposed to a total dose no greater than that which was received in only two weeks by the Apollo 12 astronauts.

Occasionally, for reasons we are only slowly coming to understand, the Sun emits sudden bursts of radiation called flares. These rays travel almost as fast as light, and reach Earth within minutes. When they do, they cause brilliant auroral displays in the upper reaches of our atmosphere. Very rarely, every few decades, particularly intense flares occur, which saturate Earth with radiation, temporarily blank out much of our long distance radio communications, and even affect Earth's magnetic field. Such an event last occurred in the 1950's. If there had been astronauts on their way to the Moon at that time, they would almost surely have been killed by that flare. Therefore, even the first space community must be protected against solar flares and heavy primaries. This could be done by passive shielding, using lunar surface material or the slag from the industries of the early colonies. The thickness required would be some fifty centimeters (twenty inches) of sand or its equivalent. That would be enough to increase noticeably the required mass of Island One.

The effect of that shield thickness, oddly enough, would be to enhance to an unacceptable level the radiation from the galactic primary rays. The reason is that on encountering dense matter those particles would break up into many more, of lower average energy but much greater total numbers.

In the end, then, we must do the entire job and get rid of all three components of radiation. When the numbers are worked out, we find that the shielding needed is substantial — equivalent to about two meters (over six feet) of soil. Once that problem is thoroughly understood, it constitutes a serious restriction on the design of the first habitats. Fortunately, a geometry has been found that fully satisfies even the most severe shielding requirement, without sacrifice of desirable design features.

The later space communities, of the size of Island Three or larger, will have atmospheric depths and thicknesses of structure below the ground great enough that they will afford to their inhabitants protection from cosmic rays comparable to that of Earth. Their building materials, the lunar soils, are already known to be fairly similar to those of Earth in natural radioactivity.[6]

To summarize, with proper design, both the early and the later space communities can be shielded against all types of radiation to levels comparable with what is found here at the surface of the Earth.

In order to minimize costs, probably the early habitats will have atmospheres composed mainly of the material most plentiful on the Moon: oxygen. The National Aeronautics and Space Administration has reason, though, to be apprehensive about pure oxygen atmospheres. In 1967 three prospective Apollo astronauts died in a flash fire in an Apollo module at Cape Kennedy, during a test conducted in pure oxygen.

The conditions of a space community will be different in several ways. First, the oxygen pressure will be only one fifth as high. At the Cape in 1967 the disastrous test was conducted with oxygen at the full sea level pressure which is normally made up mostly of inert nitrogen. Second, the volume of a habitat will be millions of times larger than that of an Apollo module, so that any small fire which starts within it cannot build up the gas pressures which were destructive in the Apollo test.

Possibly, though, these two differences will not be enough. To be on the safe side we want an additional security factor. One approach is to add a special component to the atmosphere, something that is harmless to humans but that either would not support combustion, or would actively damp it. We should first consider obtaining a damping gas from lunar materials. On Earth, fires are partially damped by the presence in our atmosphere of nitrogen. The lunar surface materials are known to contain small amounts of volatile gases, so that in processing a million tons of lunar materials a few thousand tons of gases will be evolved.

Their composition is not as accurately known as we would like, but it is thought to be mainly carbon dioxide, nitrous oxide and a small percentage of water. We might be able to get a useful amount of nitrogen from that source. It doesn't seem likely, however, that nitrogen will be a very effective fire retardant. Even if we find a cheap source for adequate quantities of it (which seems unlikely), we cannot put much nitrogen into the space community atmosphere without raising the pressure enough to increase the habitat structural requirements.

There are gases which are harmless to humans at least for short times, but which actively retard fires; some of the freons have this property. But these are chemicals made of elements not all of which are found on the Moon, and we lack adequate data on their long-term physiological effects.

It appears now that the simplest solution would also be the best. To maximize the day-to-day pleasures of life in the space colonies, as well as their safety, it seems wisest to bring along from Earth enough hydrogen so that the atmosphere will have a comfortable relative humidity, and so that there will be plenty of lush green vegetation. Structures there will be made of non-burning materials, similar to brick or cinder block on Earth, so with a combination of reduced atmospheric pressure, large total volume, and plenty of water the fire danger appears reducible to an acceptable level. This is an area in which actual laboratory research here on Earth will be required before the answers are certain.

With regard to war we must be speculative. I hesitate to claim for the humanization of space the ability to solve one of mankind's oldest and most agonizing problems: the pain and destruction caused by territorial wars. Cynics are sure that mankind will always choose savagery even when territorial pressures are much reduced. Certainly the maniacal wars of conquest have not been basically territorial. When Genghis Khan conquered most of Europe and Asia he had no plan in mind for the conquered lands, and therefore simply destroyed their cities and murdered their people. Yet the history of the years since the second world war suggests some changes relative to the past. If anything, warfare in the nuclear age has been strongly, although not wholly, motivated by territorial conflicts — battles over limited, non-extendable pieces of land. It appears that the territorial drive to conquer someone else's land should be muted under the conditions of the space communities. They will be free of the age-old associations which fuel territorial wars on Earth, they will be replicable so that no one need feel constrained by a fixed boundary, they will be independent of each other for their essential needs, and they will be movable. In the long run, when new habitats may be built most economically at the asteroids themselves, upon completion their residents will have a choice: to move, by low thrust engines over a period of decades, to an area in which other, culturally congenial communities are already located, or — go the other way.

From the viewpoint of international arms control, two reasons for hope come to mind. We already have an international treaty banning nuclear weapons from space, and the space communities can obtain all the energy they could ever need from clean solar power. The temptations presented by nuclear reactor by-products need never exist in space.

From the viewpoint of a military man, the space habitats will seem rather unpromising as sites for weapons or military bases. First of all they will be quite vulnerable militarily, so that no one in such a habitat can be tempted into believing that he can attack someone else without risk to himself. Second, their distance from Earth, and their consequent separation from it by at least one or two days of travel time, will mean that they can never be used as effective sites for an attack on the home planet. In summary, the probability of wars between the habitats seems, to me at least, considerably smaller than that of wars between nations on Earth.

At lectures on space communities, an occasional question concerns the possibility of attack on a colony from within, by some insane person or extremist group bent on mutual annihilation. The possibility is there, at some level, but probably it will carry with it some safeguards of its own. I suspect that many habitats may choose to have some sort of "customs inspection" which would eliminate or greatly reduce the likelihood that explosives or weapons could be introduced into them. In the past years on Earth we have come to take inspections of this kind as a matter of course at all airports. If, in spite of such precautions, a terrorist were somehow to import or manufacture explosives, he would have to do so on a fairly large scale to produce a major catastrophe. Like airplanes, bridges, and ships, the habitats will be designed so that loss of a single supporting band, or of a single longitudinal cable, will not result in a major rupture but only in the redistribution of loads to the supporting members nearby. As discussed earlier, the destruction of one or even several window panels would result only in a loss of atmosphere slow enough that there would be plenty of time for evacuation to communities nearby.

The external tension and compression towers, which may provide for each cylinder the forces necessary for its precession about the Sun, would not be very vulnerable to terrorists, located as they would be in space where no one could move without a space suit. If, though, one of them were to be destroyed, either by accident or by intention, it wouldn't result in catastrophe to the habitat. The precession would be arrested, so if repairs took as much as a day the residents would see the image of the Sun's disc wobbling by about two solar diameters, though the intensity of sunlight would be undiminished. On completion of repairs the precession rate could be speeded up to a rate greater than normal, until the community "caught up" to the correct orientation. Such an event would be seriously damaging only if repairs took more than one or two weeks, so that Sun angles were changed by many degrees and crop growth was correspondingly affected.

Certain dangers exist on Earth but would not in a space habitat. Earthquakes and volcanoes are among these. Often they wipe out thousands of people at a time, particularly in seacoast areas. Tornadoes, hurricanes, and typhoons also kill, and numbers of people are killed every year in small boat accidents through weather or violent waves. Among the risks which our technical society has added are those of automobile accidents. Because of good roads, safe automobiles, and relatively strict traffic laws, in the United States we have about the lowest accident rate per passenger-mile that is found anywhere in the world. Yet even our rate results in the death of 50,000 people per year, out of a population of two hundred million. One comparison between the risks on Earth and those in a space habitat is instructive: even in the extreme case in which it is assumed that a meteoroid strike of one ton size on a space habitat would result in total destruction and the loss of all the inhabitants, the risk of death from that cause would be only one sixtieth of that which we run in the United States by the existence of our automobiles.

If the space habitat option is followed on the earliest possible time scale, the result could be that within a few decades the nations of the world would all be dependent on solar energy from satellite solar power stations built at space communities. Nuclear energy, under those conditions, would be confined mainly to the laboratory. Dependence on a relatively vulnerable but inexhaustible power source would remove one of our present causes of international tension and the threat of war, and at the same time would deter any would-be adventurer nation from carrying out an attack on a neighbor.

In contrast, if for our energy we are forced to rely on a rapid, large scale development of liquid metal fast breeder reactors, within a few decades every industrial nation and every developing nation will have such devices. Plutonium will be in production in large quantities in every such nation, and the temptation to divert it to weapons production will be very strong for at least some political leaders. With so much fissionable material being produced and shipped, it seems likely indeed that some of it will be diverted by terrorist groups, and consequently Earth may become a much more dangerous place than it is now.[7]

In terms of risk, therefore, the alternative appears to lie between a development of space communities, relatively safe from catastrophe, in which an increasing fraction of the human race would be widely dispersed and consequently safe from simultaneous destruction, and an Earth ever more crowded with population, on a strictly limited land area, under conditions in which the probabilities both of war and of terrorist acts would be enhanced.

# Chapter 8
# The First New World

The first space community large enough to form a powerful industrial base, able to manufacture products of value in quantities great enough to provide important economic benefits to Earth, will require a population of at least several thousand people. A space station supporting only a few astronauts would be far too small to "seed" the manufacturing program. To build a sizable habitat will require making full use of the advantages of scale. The experience of space exploration so far is that the development costs of new vehicle systems tend to be underestimated, while the economies of scale, of quantity, and of size tend to be insufficiently used. Beyond a certain lower limit, the cost of greater tonnage transported to orbit is only that of additional launch operations, which become less expensive as once-developed systems are replicated and progress is made along the learning curve for their construction. For that reason it wouldn't cost ten times as much to establish in space a work force ten times as large. We can't say for certain what will be the minimum number of people needed in space in order to reach the "ignition point" — the level where they will be generating new wealth fast enough so that further growth won't require subsidy from the Earth — but all the studies made so far agree that ignition will be reached by the time the population in space reaches 10,000. If those people are only as productive as an equal number engaged in heavy industry on Earth, their output every year of finished products will be more than the mass of several ocean liners.

Concentrating on the "nuts and bolts" details of the construction of Island One, we must keep clearly in mind the difference between science fiction and reality. That difference is the contrast between practical technology and unchecked imagination. We must depend only on present day technology, on machines which we are sure we can build within the limits of our present knowledge, and on costs calculated with as much realism as we can attain. Time scale is of the greatest importance. Unless Island One can be built rather quickly, its productivity will be of no use to us in the time scale for which we may be willing to commit investment. That constrains us to what the professionals call "near-term" launch vehicle systems. In our design work we must restrict ourselves correspondingly to practical engineering and sensible economics.

The Island One design is a Bernal sphere with a central sphere with a diameter of about 500 yards at which Earth normal gravity exists.
At each end a banded torus contains the agricultural areas.
An Island One colony would comfortably accommodate 10,000 residents.

In the early days of every remote construction project, accommodations are modest and living conditions are rather simple; the amenities come later. We've seen that history repeated with the construction of the transcontinental railway in the last century, and with the opening of the Arabian oil fields during the past few decades. By the time the population in space passes the ignition point, though, we can expect that pressures will be strong to transfer from modular apartment-like habitats to something large and Earth-like. I'll give now an "existence proof," a demonstration that one possible, workable design exists to house and support 10,000 people in comfort and safety. No one will be more surprised than I if, when Island One is completed, it looks very much like the sketches we now make of it. Even its size and its population may be quite different. If we go by the almost universal human experience of large scale construction projects, it will probably end up being smaller, and costing more, than our first estimates indicate. Knowing that from the start, we should take care to develop the design of Island One in such a way that it can be reduced in dimensions, or as engineers say, "de-scoped" as the design progresses.

For conservatively chosen figures on agricultural productivity, we'll need a growing area about equal to a square 0.8 kilometers on a side. There will be no need for that growing area to be spacious or beautiful; an agricultural plant doesn't care whether it has an open sky above, or only a ceiling. Sunshine in great quantity will be required, as will water, soil and nitrates.

Plants are relatively insensitive to radiation, so there appears to be no need to provide the agricultural areas with radiation protection. In the early days, though, before we have sufficient

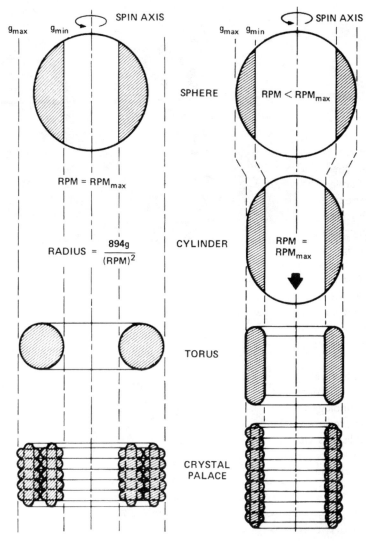

$$RADIUS = \frac{894g}{(RPM)^2}$$

A sphere doesn't provide quite the maximum usable surface area per unit volume, but the Bernal sphere design represents a minimum risk project for a first habitat when considering structural mechanics and distributed functionality.

experience, it may be wise to grow our seed crops within the living habitat where full protection from cosmic rays and solar flares will be provided.

One quite efficient design for agricultural areas consists of a series of partial wheels (tori) connected to form large fields all at the same level. Planting and harvesting machines as large as the biggest combines ever seen in the wheat fields of the western plains can move freely over those fields without obstruction. Sunshine will enter through glass windows, and the appearance will be not unlike that of a large greenhouse. In comparison with the alternatives, this design will use so little structural mass that the efficiency of agricultural productivity will become unimportant. If, after additional research, it is found necessary to double the area allocated to agriculture, that change will add very little to the total structural mass of Island One.

More than a century ago Queen Victoria's consort, Prince Albert, led a distinguished group of British industrialists in the design and realization of the International Exhibition of 1851. The central feature of the exhibition was the Crystal Palace, a light and airy structure made of glass windows set in modular ironwork. So light and so well designed was the Crystal Palace that it was assembled within a few months by a construction crew of quite moderate size. A whole avenue of trees and acres of exhibition space were

enclosed. Our multiple torus geometry for agricultural areas strikingly resembles the Crystal Palace, with its arching vaults of glass.

As in the case of high yield agriculture on Earth, most farming activities will be mechanized, so radiation shielding for the operators of tractors and combines can be incorporated into the machines themselves.

Light industry, of benchtop scale, may be carried on within the living habitat, but heavy industry can make use of the zero gravity of free space.

The modular (Crystal Palace) approach offers maximum construction efficiency because of the repetitive nature of the manufacture and assembly operations. Much of the production and labor can be automated because it is repetitious.

The design requirements for the living habitat in Island One are severe. The habitat must admit sunshine easily, yet be fully shielded from cosmic radiation. It must provide a spacious, comfortable environment, with long sight-lines to prevent the inhabitants' from suffering claustrophobia. Ideally, it should provide easy access to a region, fully shielded, where zero gravity sports can be enjoyed. For safety, mechanized transport should not be relied on — in the event of a sudden major emergency, it should be possible for the entire population to move rapidly, without mechanical assistance, to docking ports for evacuation. Finally, the habitat must be economical of mass, both in structure and shielding.

The land area of a habitat for 10,000 people can be estimated from considerations of "personal space" and the experience of small towns on Earth. A typical garden apartment community in an affluent section of the United States provides, with its swimming pools, tennis courts, and landscaping, about 45m$^2$ of total land area per person. For comparison, the city of San Francisco, averaging over both residential and park areas, provides about twice that area for its population. Some of the attractive hill towns in southern France and Italy have only about one fifth as much.

One possible geometry that satisfies all these requirements is simple and structurally strong — a sphere one mile in circumference, with sunshine brought inside through windows. If the sphere rotates twice per minute, it will provide Earth-normal gravity at its equator, near which most apartment areas can be located. At the forty five degree "lines of latitude" halfway up the inner surface of the sphere from the equator, gravity will be a third less than Earth-normal. That variation from Earth conditions may be our self-imposed "design limit" until we gain experience on physiological tolerances.

In such an environment each family of five people can enjoy a private apartment as large as a spacious house (230 square meters of floor area) with a private, sunlit garden of a quarter that area. By arranging the apartments in terrace fashion, only a small fraction of the total spherical surface area below forty five degree latitude need be devoted to apartment gardens, most of the remainder being available for parks, shops, small groves of trees, streams, and other areas available to all inhabitants.

The sunshine will enter, during a day length set by the settlers' choice, always at a fixed angle. That will permit providing every room of every apartment with natural sunshine throughout the day. On Earth, a narrow aperture between buildings can receive sunshine only for a few minutes each day, but not so in space, where each window may look out onto a sunlit, private mini-garden.

The equator seems an ideal location for a wandering, shallow river, opening into occasional deep pools for swimming. The shoreline beaches can be of lunar sand, and perhaps at a little distance, surrounded by greenery, there can be paths for bicycling, walking and running.

When structural details are examined, it develops that the optimum location for windows will be near the rotation axis. There, only the pressure load will be important, and gravity will add little to the structural demands. The sphere will be no fragile eggshell, though. Its aluminum wall will equal the thickness of battleship armor, up to seven inches at the equator.

Low gravity swimming pools and "hangars" for human powered aircraft can be located near the rotation axis. Walking to them from the equator will be equivalent to climbing a gentle hill, and should take only about twenty minutes.

For a given volume enclosed, a sphere is the shape that requires the least surface area. That is important for minimizing the required mass of cosmic ray shielding. For economy, the shield can be made of unworked lunar soil or industrial slag packed between thin spherical shells spaced a few meters away from the rotating habitat. It is possible in such a geometry to bring natural sunshine into the habitat through mirrors all of which are stationary in space. Only much later in the history of space communities need the designers concern themselves with such complications as rotating mirrors.

With complete shielding, provision must be made to remove from the living habitat the heat brought inside by sunlight. One easy way appears to be through large axial passageways divided by a cylindrical shell. Air circulation through these passageways will remove the heat to external radiators, and the same corridors will serve for the zero gravity movement of people and freight to and from the industries and docks outside.

If desirable, it will be rather easy to separate the sphere visually into three "villages." That arrangement will permit making the day length and time of day of each village independent of all the others. That in turn will allow a convenience and source of efficiency forever denied us on Earth: In order to get the most out of machines, chemical processing plants, and other industrial facilities, they should be run full time. On Earth, in order to do that we must subject people to working night shifts, which almost no one likes. In Island One, though, three villages can run at time zones separated by intervals of eight hours, so that industries can run full time while everyone remains on his own "day shift."

For structural simplicity, we want to avoid in our design any rotating pressure seals. The habitat should rotate as a unit, airflow being contained within a single pressure vessel. Combining the Crystal Palace geometry for the agricultural areas with a central sphere for people, we arrive at the design concept called Island One.

The structural mass of Island One has been checked by calculations in several studies, and is about equal to that of a large ocean liner like the Queen Elizabeth II, 100,000 tons. Buildings, soil and atmosphere will be several times as much, and even in this most efficient design shielding will add another three million tons.

To summarize, Island One will be small, though far less crowded than many Earth cities, and it can be attractive to live in. The inhabitants can have apartments which will be palatial by the standards of most of the world. Each apartment will have a private garden, bathed every day in sunshine at an angle which will correspond to late morning. Even within the limits of Island One and its water supply the colonists can have beaches and a river, quite large enough for swimming and canoeing. The river will offer a possibility that some people will be sure to exploit — a floating trip, past the dam, filters, pump area and spillway that interrupts the circular river at one point, all the way around the cylinder circumference to the starting point.

Even within Island One the new options of human powered flight and of low gravity swimming and diving will be possible, and the general impression one will receive from a village will be of greenery, trees and luxuriant flowers, enhanced if the village chooses to run with the climate and plant life of Hawaii. Heavy industry can be located outside but nearby, so that no vehicle faster than a bicycle will be needed throughout the community. Island One will rotate about once every thirty one seconds, to provide Earth-normal gravity for its inhabitants whenever they are at home. Only when at work outside the habitat will the residents be

Internally, Island One can be practically a paradise – no pollution and no heavy industry, but a low population density and much recreational and park land.

subjected to zero gravity; in a daily routine of that kind their bodies will retain normal muscle tone and strength without special exercise.

The site of Island One should be far enough from Earth and Moon to avoid frequent eclipses, so the community can use free solar power continuously. We don't want it so far from the Earth that transport will be difficult, nor as close as the Van Allen radiation belts that surround the Earth. When all the logistics are considered, the best location may be simply a high circular orbit, with a period of a few days, part way out toward the Moon. There is another choice, attractive mathematically, which was studied intensively for a time. It is an eccentric orbit with a period of two weeks, just half that of the Moon. Still earlier, those of us interested in answering the question "Where will the colony be?" had considered a point on the Moon's orbit called L5, the fifth of several locations in space whose properties were first described by the French-Italian mathematician and physicist Joseph Louis Lagrange (1736-1813). In the language of the 1911 Encyclopaedia Britannica:

"He gave proof of the undiminished vigor of his powers by carrying off, in 1764, the prize offered by the Paris Academy of Sciences for the best essay on the libration of the Moon.

"His success encouraged the Academy to propose, in 1766, as a theme for competition, the hitherto unattempted theory of the Jovian system. The prize was again awarded to Lagrange, and he earned the same distinction with essays on the problem of three bodies in 1772, on the secular equation of the Moon in 1774, and in 1778 on the theory of cometary perturbations."

Lagrange used the gravitational theory developed by Newton to explore the special properties of two unique points in the orbit of Jupiter. One of these points preceded the planet in its orbit around the Sun by sixty degrees, while the other followed by the same amount. Lagrange concluded that these were in fact stable points, near which any objects with the correct initial location and velocity would stay forever. From that time on these were known as the fourth and fifth Lagrange points, described by solutions to what physicists call the restricted three body problem. Years later, observations through primitive telescopes showed that several asteroids or minor planets were trapped near the Lagrange points. These became known as the "Trojan" asteroids.

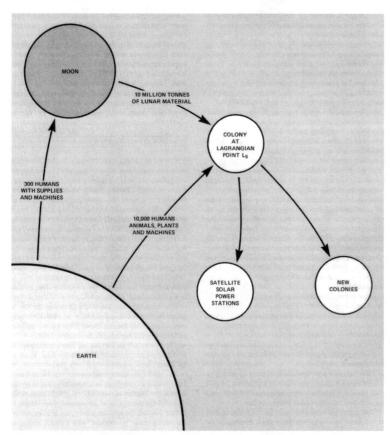

The term "L5" originally designated a fixed point in the same Earth orbit as the Moon, but 60° away from the Moon in that orbit. More recently, L5 has come to mean any stable orbit about the Earth that is far enough from both the Earth and the Moon to prevent either body from blocking sunlight (solar energy).

If we want to use the corresponding Lagrange points in the Earth-Moon system either as sites for colonies or as possible locations for useful trapped materials, we are up against a much tougher kind of mathematics. We must solve not just a three- but a four-body problem, because the Sun, distant as it is, powerfully affects orbits in the vicinity of Earth, in consequence of its enormous mass.

Fortunately, the problem has been done for us, but only just in time. In 1970, A.A. Kamel, a student of Professor John Breakwell at Stanford, obtained his doctor's degree in engineering by publishing a thesis with the forbidding title *Perturbation Theory Based on Lie Transforms and Its Application to the Stability of Motion Near Sun-Perturbed Earth-Moon Triangular Libration Points.* Dr. Kamel's work, which gives us in an elegant mathematical form a solution that had already been obtained by the brute force methods of computer calculation, tells us that in the Earth-Moon system L4 and L5 are no longer stable points, but that they are replaced by something at least as good — stable regions which move in orbits of very large dimensions about L4 and L5, on a slow, eighty nine day cycle. The properties of L4 and L5 are so unique that a society has been named after L5, and for convenience we often speak of "L5" as a nickname for "any orbit above the Earth's radiation belts, and no farther away than the Moon." It's characteristic of the orbit mechanics problem that the experts in that field often rush in, waving great stacks of computer output, and lecture the rest of us on a newly discovered orbit that's better than any found before. By now this has happened enough times that a wise man wouldn't place bets on exactly where Island One will go. The one clear message we can be sure of is that there's room in high orbit for a total population many times that of the Earth. We need not fear, by the way, that eventually our local neighborhood, the Earth-Moon system, is going to become overcrowded. Space communities could be located on orbits almost anywhere in our solar system, and with proper mirror design could still enjoy the same solar intensity which we have (on good days) here on Earth.

We can set a scale for the investment required in Island One by considering the largest space project we have so far carried out: Apollo. That venture, which will surely be remembered long after the misery and the horrors of our century have been decently laid to rest in the history books, cost about $50 billion in money of 1978 vintage, halfway between the "Apollo years" of the 1960's, and the late 1980's, which conceivably could be the years of Island One. Apollo was begun at a time when the mood of the nation was vastly different from what it is now. Then we had confidence in our abilities, we saw our living standards increasing rapidly, our money was sound and we did not yet see limits to our continued growth. Though the environment was deteriorating as a result of our industries and our transport systems, most of us were unaware of the fact.

Now these positive factors are reversed. The late 1960's and 1970's have become a time of disillusion, of slow economic growth coupled with inflation, and of living standards improving only slowly. Soon after the first Apollo landing, in 1969, we passed through a period of profound distrust of anything technological, and we will probably never again welcome new technical options in the same unthinking manner that we did in the 1950's. This is probably all to the good. Our physical power has grown so much that we should now examine with the greatest care and considerable cynicism any new technical proposal, lest it carry with it unseen dangers.

To succeed in these hard times when economic concerns are paramount, any new program must be productive — it must be able to not only pay off its initial investment but to generate new wealth. As we now see it, the first payoff from the development of space communities will be a supply of low cost electrical energy here on Earth. We should examine the scale of investment which is customary in the electric utility industry, then estimate the cost of seeding a space manufacturing program, and see whether those figures are in balance.

In 1975 the total installed generator capacity of the United States was about five hundred Gw (five hundred thousand megawatts).[1] When in 1974 the first mild energy shortage struck, a number of studies were initiated to provide estimates of how that capacity would have to grow during the next quarter century. It was assumed in most of these studies that conservation and rising energy prices would limit energy growth rates to less than 7 percent per year, which was commonplace in the 1960's.

A working group formed by the Institute of Electrical and Electronic Engineers summarized in a report twelve of these forecasts. According to its conclusions, the installed generating capacity of this country must quadruple, to about 2,000 Gw, during the last quarter of this century.[2] That increase is equivalent to an average growth of just over 5 percent per year.

In order to meet that demand for generator capacity, the electrical utilities of this country will have to spend in this quarter century about $800 billion, at a rate of $530 per kilowatt[3] The latter figure is appropriate to coal fired generating plants; nuclear plants cost considerably more. Eight hundred billion dollars is nearly as much as this country makes in a single year (its gross national product). It's almost twenty times as much as the cost of the entire Apollo Project over the decade during which that enterprise was carried out.

If indeed the establishment of a manufacturing facility in space, able to process lunar surface raw materials, can satisfy our electrical energy demands, what investment will be required to set up such a facility? By now there have been a number of independent estimates of that investment. One, progressively updated and revised, has been made by the NASA Marshall Spaceflight Center. Another, using NASA figures for launch costs but otherwise independent of the space agency, was made by a study group working in a joint program of NASA, the American Society for Engineering Education and Stanford University.[4] A year later a study group working purely under NASA sponsorship went through a still more detailed estimate of construction time and cost.

All of these estimates were based on a fairly direct approach, not yet taking full advantage of the possibilities for cost saving inherent in the idea of manufacturing in space. They agreed fairly closely, though, and centered on about $100 billion — only a small fraction of the investment that the utility industry will have to make to satisfy our electrical energy demands.

The Vehicle Assembly Building at Kennedy Space Center is where all Skylab and Shuttle lift vehicles have been assembled and prepared for liftoff. The landing runway for the Shuttle can be seen at the left.

These various estimates of investment needed agreed because no great advances in technology were assumed by any of the estimators. Once the total tonnage to be lifted into space was established, the total investment could be estimated from known launch costs and from experience on the development and administrative expenses during the first decades of our space program.

The cost, it seems, would be about twice that of the Apollo project — and we haven't yet talked about ways of further cost cutting. Although in retrospect Apollo appears as a vitally necessary prospecting expedition, essential to any serious proposal to use lunar resources, it appears that the establishment of space manufacturing would give a much greater payoff — a productive factory in space, with a self-supporting work force of 10,000 people, in contrast to a brief series of daring scientific forays by less than a dozen men. The reasons for that greater payoff are post-Apollo advances in vehicle systems, and above all the "bootstrap process" — using the material and energy resources of space to build manufacturing capacity.

We can see at once that if the materials for Island One must be brought up from Earth, there is no possibility of constructing Island One at an affordable investment cost. An Apollo rocket, costing several hundred million dollars and wholly discarded after one use, could lift payloads into orbit, but only at a cost of thousands of dollars for every kilogram. To go as far as L5 with such machinery would have cost several times more, and to haul freight to the Moon in the days of Apollo ran the cost of every kilogram as high as the price tag on an expensive sports car, some $20,000.

Even if we were to be so optimistic as to suppose that with the investment of many years of time and many billions of dollars we could develop launch vehicles able to operate at a hundredth of the cost of Apollo vintage rockets, we still couldn't afford to haul up the pieces of a space colony from the Earth. For the shielding alone, the lift costs would be a healthy slice of our gross national product. Clearly, then, to construct space manufacturing facilities mainly out of materials from the Earth would be absurd.

The Skylab Project of the early 1970's yielded a great deal of scientific and technical information, and considerably advanced our understanding of the effects of long term weightlessness on humans. Its basic rocketry, though, was that of Apollo, so it did nothing to advance the art of launching heavy payloads at lower cost. NASA is now devoting most of its development effort to a project which will push chemical rocketry to a high state of sophistication: that is the Space Shuttle program. The shuttle is a winged, orbital vehicle intended mainly for scientific missions in low Earth orbit. It's designed for reuse, at least in part, and it will be particularly suitable for missions in which scientific instruments of large size must be recovered from orbit and returned safely to Earth. In the course of developing the shuttle, NASA is putting a great effort and several billions of dollars into the design, testing and perfection of what are called "SSME's": space shuttle main engines. These are not very large engines, in comparison with those of the Saturn 5's which launched the Apollo flights, but they are a great deal more efficient. They operate at an internal pressure as high as modern materials can stand, and at temperatures close to the material limits. It will be some time before chemical rocketry pushes much beyond the performance figures that the SSME's can attain.

The shuttle is designed as a two-stage vehicle, and its first stage is a pair of solid fuel rockets which after burnout are to be soft landed by parachutes in the ocean and then (with a probability which only experience can tell us) are to be recovered and reused.

For some time now NASA has been studying designs for a freight vehicle based on shuttle engines: a "shuttle derived heavy lift vehicle" or HLV in the language of the rocketeers. It would be a booster, not necessarily manned, which could lift about a hundred metric tons to low Earth orbit. The HLV would not be a large vehicle; its height on the pad would be half or less than that of an Apollo-Saturn 5. It would be capable of higher performance than Apollo, its first stage possibly consisting of shuttle solid rocket motors and its second powered by SSME's. There are alternatives, also, to the solid rockets. Within the present stage of the art the first stage engines could be liquid fueled, burning kerosene or ammonia and liquid oxygen. Especially in the latter case, the first stage would release fewer pollutants to the atmosphere, and its fuel would cost far less than that of the solid rocket motors. Either way, the HLV could be built on a rather short time scale, taking advantage of the great effort which has already gone into the development of the SSME's.

The Space Shuttle was the original hope as the workhorse for conquering the High Frontier.

NASA is presently advertising a cost of about 20 million dollars for a shuttle launch, assuming complete recovery and reuse of all the hardware required. The SSME's would cost several million dollars each, so for economy they should be recovered from orbit. The latest HLV designs show the SSME's mounted on a re-entry shield, so that after lifting freight to orbit the engines could be recovered by atmospheric braking followed by pop-open parachutes, just as the Apollo command modules were safely recovered with the returning astronauts inside.

The Shuttle's main engines (SSME's) and External Tank could easily form the design basis for a new Heavy Lift Vehicle (HLV) capable of putting larger payloads in higher orbits than the Shuttle.

In May 1975, at a Princeton University Conference on Manufacturing Facilities in Space, two professional rocket designers with many years of experience at NASA presented their estimates for the kind of vehicle needed both to reach low orbit and to go beyond it to L5 or to the lunar surface. Hubert Davis, from the Johnson Space Center in Houston, presented data from several NASA and industry studies on HLV conceptual designs.[5] A.O. Tischler,[6] now retired after many years of service at NASA, discussed a chemically propelled "tug," an engine and control system small enough to be placed in orbit by the HLV and then capable of moving payloads of various shapes and masses from low orbit to L5. To go from lunar orbit to the lunar surface we will also need a "lander," another small vehicle quite similar to the tug. The early estimates on the investment needed for Island One and its early successors were based on just those few vehicles: the space shuttle, which made its first free flight in 1977, the shuttle derived HLV, the tug, and the lander, the last two being small chemical rocket vehicles well within the present range of engineering knowledge.

At the 1975 Princeton Conference it was confirmed that the cost of putting a ton of payload on the lunar surface would be about twice as much for the same load placed at L5, and that the cost to locate at L5 would be about the same as to place a payload in geosynchronous orbit, above a fixed point on the surface of the Earth. In later, more detailed NASA supported studies in 1976 and 1977, these estimates were checked further. Remarkably, it has been found with each successive study, as the engineering has become more complete and the cost estimation more professional, that the cost estimates for the establishment of Island One have come down.

The most recent work traced a program in which Island One would be preceded by smaller habitats, down to the size of a small space station. These habitats, the first transportable by the space shuttle, would be

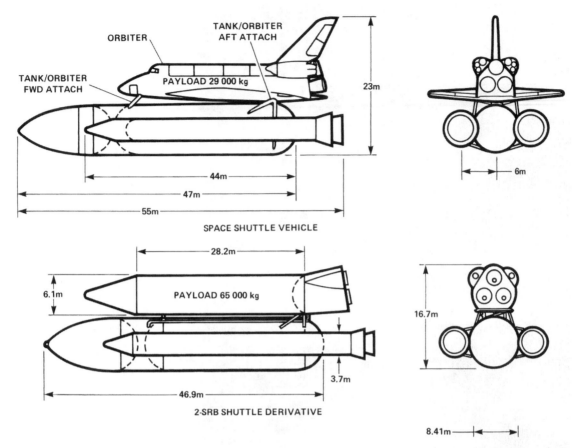

Removal of the Shuttle's wings and rudder (not needed for an HLV which does not reenter) and their associated control systems would allow for more fuel and a larger payload. If the HLV is a freight-only vehicle then all of the space and weight for life support systems and consumable can also be used.

temporary quarters for a work force whose first priority would be to set up manufacturing in space, so that the program could begin to return profits and quickly pay off the investment made in it. Only after the program was solidly established on a paying basis would the productivity available in space be diverted even in part to the construction of something as luxurious as Island One. In that scenario, it might be one or two decades after the initiation of space manufacturing before Island One and its counterparts would be completed. Apparently, by adopting such an approach the investment required to reach the "ignition point," after which the profits from space manufacturing would sustain further growth, would be cut to only a small fraction of the amount necessary for the construction of Island One as an initial project.

By now we see clearly, I believe, the logical building blocks in our program of space manufacturing. We can put them together in different ways, and in order to get the greatest payoff for the least investment we'll be studying all the possibilities right up to the moment when the final planning decisions have to be made. Let's look at those building blocks one by one, though, because they're likely to turn up in any final program plan.

At the 1975 Princeton Conference and at the Summer Study of the same year "refueling" calculations were made. These indicated that when liquid oxygen derived from lunar materials is available at L5, both the cost and the number of launches required from the Earth can be reduced greatly. In fact, when oxygen from the industrial activities at L5 does become available, it will so greatly reduce the cost of tug operations that the chemical tug will perform at a level otherwise unobtainable except from an advanced nuclear rocket. This fact may dictate that the first industry processing lunar materials extract the oxygen alone. The potential savings from that method have not been put into the cost calculations made so far.

A lunar base with a mass driver would provide continuous raw materials and would require only a very small permanent staff to operate and maintain.

## LUNAR SOIL COMPOSITION

Aluminum: 7%
Magnesium: 6%
Other: 3%
Calcium: 8%
Iron: 13%
Oxygen: 42%
Silicon: 21%

The lunar soil is known to contain many of the elements needed for space construction, life support and rocket propellant.

The idea of using lunar oxygen for chemical rockets isn't new, by the way. Robert Goddard thought of it a half century ago, and Arthur Clarke brought up the same idea some years later.

When we look into the economics of space manufac-turing, we find that over a few years several million tons of lunar material must be processed. To keep the investment cost down and to keep the number of shuttle and HLV flights within NASA's "traffic model," though, we'd like to hold the lunar installation to not more than a few thousand tons.

The installation on the Moon must therefore be able to launch during a few years a thousand times its own weight. No rocket within present technology could achieve such a figure. We must design instead a transport device that can launch payloads from the Moon without itself ever leaving the surface.

Before we go into the details of the transporter, we should consider how the "bootstrap" principle of establishing a launcher on the Moon can yield a growth of space habitats and of their products without further drain on the resources of Earth. Clearly the first such launcher must be built on Earth, tested and perfected here, and then launched to the Moon and reassembled there. By its presence it will then permit the construction of the first space manufacturing facility at an affordable cost. Once the first habitat is in place at L5, one of its first products, logically, will be additional transporter devices. The cost of moving them from L5 to the Moon will be substantially less than that of bringing more transporters from Earth, and as the total installed cost will be dominated by that of transportation, Island One will become the favored location for their production.

In order to rid ourselves of what Isaac Asimov calls our "planetary chauvinism," we should consider why the Moon, though it is necessary as a materials source, is less suitable than L5 as a site for industry and human habitation. We can be rather quantitative about some of the reasons.

First, the cost of transporting workers and their families to the Moon, and the cost of transporting from Earth the necessary machines and tools, liquid hydrogen, chemical processing plants, and an initial construction station large enough to build a habitat, would all be roughly twice as high as for transport from Earth to L5, so the amortization cost of all such equipment and materials would be far higher on the Moon than at L5. In turn that would increase the price of any products of lunar industry.

Second, any objects which the Moon could build would then have to be lifted off by rocket power. That would limit them to comparatively small sizes. In contrast, the L5 communities could build objects that mass up to tens of thousands of tons, could assemble and test them in their final form, and could then move them to any free space location where they would be used. Lift costs by rocket from the Moon would be many times higher than the transport costs by mass driver of the corresponding raw materials.

Third, all the construction efficiencies at L5 which I have described depend on the availability there of constant, dependable solar power, for all energy needs. On the Moon, solar power would be turned off for two weeks out of every four. Though ultimately it will be possible to obtain electric power at any point on the Moon from power lines drawing from solar stations on the "day" side, electric power on the Moon will necessarily be more expensive than at L5, because on the Moon one will have to build two or three solar stations to obtain constant electric power, instead of

An electromagnetic mass driver, powered either by solar energy or a nuclear reactor, can continuously put raw materials into orbit where they can be captured and processed.

just one. The problem of supplying the equivalent of sunshine for agriculture, and heat for chemical processing, during the lunar night, will increase further the costs of operations on the Moon.

Gravity on the Moon is a problem for several reasons. It cannot be turned off, so all the possibilities of containerless processing, the building of large fragile structures, high purity zone melting, and the other attractions of zero gravity are forever denied to lunar industry.

The inescapable lunar gravity poses a further problem for any large work force located there — it's too small to keep muscles and bone in good condition without strict exercise, and yet it's enough to prevent easily obtaining one gravity by rotation. In free space, for a habitat of modest size, the cost of rotation to imitate Earth's gravity would be only a small addition to the cost of enclosing an atmosphere. Yet on the Moon to accomplish the same result we would have to build a relatively heavy structure supported on massive bearings.

When we consider that any lunar employee will have to put up with no sunshine, or artificial sunshine, for two weeks out of every four, that his transportation cost to the lunar surface will be about twice as much from the Earth, and that he will probably have to spend a considerable amount of time in hard exercise to avoid losing muscle tone, we can see that industry on the Moon will have a difficult time competing with industry at L5. It will have advantages only for such specialized products as mass drivers and their solar power plants. The Moon seems, therefore, likely to remain an "outpost in space," similar in some respects to Antarctic scientific colonies.

In the long run, as the communities continue to grow in numbers and size, presumably the lunar station will grow also. For nearly all products it will be unable to compete economically with the L5 facilities because of its permanent disadvantages of intermittent solar power, confinement to a non-zero gravity for construction, and greater remoteness in terms of rocket transport. It will have a great advantage for just one class of products: those whose end use will be on the Moon. Probably the first of these products will be the transporters, and the second may well be solar power plants for local use. In the long run, it seems logical to assume that solar power stations will be located at several points around the lunar circumference, linked by transmission cables, to provide solar electric energy without interruption. There may also be a possibility of locating stations on a high peak near one of the lunar poles, where sunlight would be available more nearly full time. All such possibilities are, though, for a later period. At first the lunar operation will presumably be confined to a single location, from which the miners and engineers will never stray very far.

The initial lunar mining setup will most likely be comprised of habitats, a mass driver and processing equipment prefabricated on Earth.

As the economic picture grows, the reader will see that the success or failure of the entire space manufacturing concept rests on the bootstrap principle, and therefore, on the transport device that must transfer lunar materials to the processing plant and industrial site at L5.

For convenience I call this device a "mass driver." As presently conceived[7], it's a kind of recirculating conveyor belt. By the action of magnetic impulses driven by electrical energy, it can accelerate a small "bucket," containing a payload of compacted lunar material, to the lunar escape speed of 2.4 kilometers per second. Then, after final guidance and precise correction of errors in direction and speed, the bucket will release the payload, slow down to a relatively low speed, and be returned to pick up another payload. The key feature in such a method is that nothing expensive will ever be thrown away. A bucket can be extremely costly, and yet will contribute little to the costs of launching. As the numbers work out, each small bucket will be re-used every couple of minutes. Even if each one were to cost as much as a million dollars, that cost amortized over a few years would add only pennies per kilogram to the cost of launching lunar material into space.

The mass driver is a device which could well have been imagined a century ago, as soon as physicists had achieved a good understanding of electromagnetic fields. An early variant of it is described in a publication fully twenty five years old, by that dean of science fiction writers (and at that time active working scientist) Arthur C. Clarke.[8] In the *Journal of the British Interplanetary Society* Clarke worked out the basic mechanics of electromagnetic launch from the Moon, and compared the problem to military research then in progress on electromagnetic launching of aircraft from carriers.

Three developments have brought the mass driver from the realm of science fiction to that of possible practicality. The first is the notion of recirculating buckets. That could have been worked out at any time, and I am still searching for evidence that someone may have written it down many years ago in some publication not yet known to me. The second is the development, just within the past decade, of superconducting wire in commercial quantities. Only now is it possible to build a magnet out of superconductor, and have that magnet operate continuously with a high magnetic field in the absence of a power supply. For the buckets, the superconducting coil will constitute a "handle for the bucket," because it will set up a constant current which external pulsed magnetic fields can grab.

The third necessary development is a curious one. It would have been possible many years ago to accelerate an object by magnetic fields, but for the lunar launch problem the difficulty was how to guide it. At the necessary high speeds, wheels would fly apart; frictional contact would waste too much energy and generate unwanted heat. The solution lies in an idea first published by a French engineer, Emil Bachelet, more than sixty years ago. That concept, "dynamic magnetic levitation," consists in the observation that if a permanent magnet moves rapidly near a conducting guideway (which can itself be a simple, curved aluminum trough), its magnetic fields generate induced currents within the guideway.[9] Those currents in turn produce magnetic fields, which act to repel the magnet and so produce a lifting force. The higher the speed, the more the lift and the lower the drag. Within the past few years design studies of this concept have reached a fairly large scale, with model "magneplane" guideway systems in operation in several countries. The magneplane or "electromagnetic flight" concept has arrived at just the right time for use in the mass driver.

If we follow the construction of the mass driver, we may see it in spectacular operation under test on the Earth. It will be a slim, lightweight tube surrounded by coils, no bigger around than a dinner plate, but many kilometers in length. At intervals there will be small capacitors for the storage of electrical energy, and every coil will be connected to a transistor-like solid state device to pulse that energy into the coil as the bucket goes by.

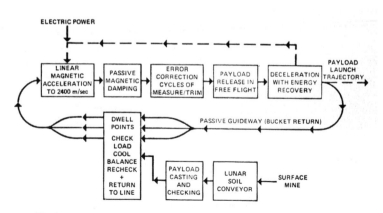

The lunar mass driver will be a "recirculating" type in which "buckets" are used to fire small payloads and then are returned on the track for further use.

We may view the mass driver only through a window, because it will be designed to operate in the near perfect vacuum of the lunar surface, and here on Earth can receive its final tests only in a vacuum chamber. Near the "injection" end a bucket will slow to a halt for a fraction of a second; a mechanical conveyor belt will remove it from the guideway for checking, automated inspection, reloading with another payload, and balancing. In its place the conveyor will set another, preloaded bucket. The first accelerating coil will pulse, and then as the bucket passes through each successive coil it will interrupt a light beam, to trigger that coil and push the bucket to a slightly higher speed. The interrupted light beam principle is the same one that's been used for many years to hold open doors as people enter elevators. When the bucket reaches full speed it will slow a little to release the payload, then will be deflected, will be braked rapidly by decelerating coils, and when slow enough will go round a gentle curve and be returned at moderate speed to the starting point. Its payload and those that follow will subject the "catcher" to a steady battering that will average to a force of more than four tons.

To supply the lunar mass driver with electricity, the alternatives are a solar cell array or a small nuclear plant. We won't need a great deal of power — only about a tenth as much as a typical generator in a power plant on Earth. The latest studies indicate that a solar cell array will be so much lighter than a nuclear plant that solar power will be preferable, even though it can only operate during the lunar daytime. As far as can be seen at present, that is the only place in the entire space manufacturing concept where nuclear power can even come close to being cost effective.

Like its cousins, the particle linear accelerators used in high energy physics laboratories on Earth, the mass driver can still work even if some of its coils fail to operate. We plan to add extra coils along the length of the machine, and in normal operation those coils will be turned off, sitting quietly as spares. In case of a component failure one or more of the spares can be switched on, so that the mass driver can continue to operate with high reliability. In the maintenance period, probably during the lunar night, the repair crew will go over the machine and replace anything that has gone bad.

In the whole space manufacturing concept, everything except the mass driver is a variation of something we've done before. The rockets are conventional, and the manufacturing operations are novel because of their location in space, but are otherwise at least analogous to bridge building and other operations on Earth. The space habitats are unique in shape because of their use in vacuum and in zero gravity, but otherwise have their analogs in ship building, aircraft and the construction industry. No one's ever built a mass driver, though, and because of that we have to work through all the basic theory of that machine, and make working models at each stage of development in order to be sure that our thinking isn't going astray.

After I published an article including the mass driver concept, in 1974, little was done to explore it more thoroughly until 1976, when I led a NASA supported study investigating possible "show stoppers" in the space manufacturing concept. In that study I had the great good fortune to work with Dr. Henry Kolm, of M.I.T., and Dr. Frank Chilton, of Science Applications in California. Kolm and Chilton had been the leaders of groups applying the ideas of magnetic flight and linear electric motors to new possibilities for high speed ground transit systems. Their groups had developed successful working models, as well as a great deal of basic theory organized in published reports and articles. It's a sad commentary on the decline of an American sense of drive and courage that both projects were killed by the government's Office of Management and the Budget during the early 1970's. At that time leadership passed to Japan and Germany, and with over a hundred million dollars being spent every year by each of those countries on magnetic flight research, by 1977 there were full scale magnetically flown test vehicles in operation in both. If belatedly we decide we need magnetic flight to solve our rapid transit problems in the United States, we'll then have to spend our dollars abroad, with unfortunate effects on our balance of payments, to buy back the developed technology that we could have had for ourselves if we'd been wiser.

With the professional experience and expertise of Kolm and Chilton brought to bear on the problem, in 1976 we were able to answer the most important single question about mass drivers: Was the idea fundamentally sound and practical? Both experts were quite sure the answer was yes. Kolm suggested that we switch to an "axial" geometry for mass drivers. In the axial case all the coils would be circular, and the drive forces could be higher. Both men were sure that my old calculations on bucket acceleration were far too conservative. In their estimation we could achieve accelerations of several hundred gravities, shortening the length of the mass driver accelerators.

In late 1976 and early 1977 I was able to devote a great deal of time to mass driver research, under the best possible circumstances. I was on sabbatical leave from Princeton, and had accepted an invitation kindly extended by M.I.T. to become the Hunsaker Professor of Aerospace for that year. It was a great opportunity for close cooperation with Henry Kolm, and we worked together throughout the year, our locations at M.I.T. being only a block apart.

My main effort was on mass driver theory. In completing the articles from the 1976 NASA study I worked out the optimization of masses, and so learned how best to design a mass driver in order to get the highest performance with the lowest possible weight.

During the first semester of 1977, at the invitation of Professor Rene Miller, Chairman of the Aerospace Department at M.I.T. and President of the American Institute of Aeronautics and Astronautics, I gave a series of seminars exploring the questions of acceleration, guidance, design and applications. Those seminars formed the basis of a 1977 NASA supported summer study task group effort, in which Henry Kolm, Stewart Bowen, several excellent students and I worked together to put the seminar results in the form of usable computer programs, and to further extend our knowledge as far as possible.

Meanwhile, we reached an exciting new stage in understanding — the building of the first working model. In the winter of 1976-77 Kolm and I designed an axial mass driver about as long as a cross country ski. The comparison is appropriate for another reason: that winter turned out to be the hardest in living memory, and my recollection of it is of great quantities of snow and ice. We had no construction budget until months later, so in January 1977 we enlisted the unpaid volunteer help of a number of students,[10] and of a young post-doc, Bill Wheaton. Our materials were from the scrap pile in Kolm's laboratory, supplemented by odds and ends like copper plumbing pipe, the brushes from an automobile starter motor, and capacitors of the kind photographers use in their flash guns.

By early May the model was done, and we demonstrated it at the last of the seminars. Then it was brought to Princeton, and for the next several months traveled quite a bit. At Princeton it became the star performer at a large conference, and was photographed in action by several television crews. Then it was shipped to California, and climaxed the final briefing of a 1977 NASA study on space manufacturing, at the Ames Research Center. From there it traveled to Los Angeles and performed (flawlessly) before an audience of a thousand people, invited by Governor Jerry Brown to a celebration of "California in the Aerospace Age," on the day before the first free flight of the space shuttle orbiter.

In the model, the bucket accelerated from zero to eighty miles an hour in a tenth of a second. Significantly, the acceleration in that first model was already higher than my estimates of several years earlier for an "ultimate" lunar launcher. Through all these travels two students, Kevin Fine and Bill Snow, carried out the setup and operation tasks. Later in 1977 Kevin continued the work and completed a master's thesis on the subject of mass drivers.

By then, a modest amount of NASA support for research and development of mass drivers was available, and we began a joint program at Princeton and M.I.T. to build a high acceleration model. By the beginning of 1977 I felt confident enough in our understanding of mass drivers, and the calculated performance figures had improved so much, that I could apply the concept not only to the lunar materials launcher but elsewhere in an updated version of a lowest cost, maximum payback plan for space manufacturing (more on that a little later). Now let's trace the flow of material from the Moon to and through a processing facility in space.

Lunar mining need not be a large scale operation. Chemical processing can be done at L5, and all industrial slag produced there will be usable as a matrix for crop growth, as shielding against cosmic rays, or as reaction mass for mass driver engines in free space. For that reason there will be no need for initial separation of the lunar surface material by high temperature processing. Experts in commercial ore processing who studied the problem believe that it will pay to "beneficiate" the lunar material, carrying out separation by sieving or magnetic effects, to increase the fraction of useful elements. After those basic operations the material can be compacted, bagged and prepared for shipment.

Dr. David Criswell has studied the problem of containing the lunar material during its travel from the Moon into space, and has worked out the details of a facility on the Moon that would make glass fibers and weave them into bags for the material. Fortunately, typical lunar sites have large quantities of glass lying about, in the form of sand that can be melted by solar furnaces.

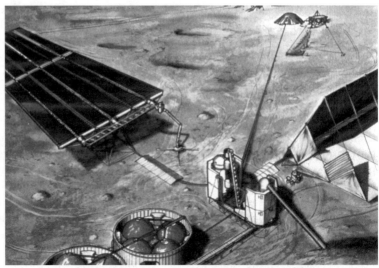

Lunar mining equipment and transport of raw materials to the mass driver site will be powered by solar energy.

When one first hears the phrase "mining the Moon" one thinks in terms of vast open pits, scores of giant machines, and a scale of operations comparable to our great terrestrial mines. The reality will be far more modest. If the surface is excavated even to the depth of a shallow gravel pit, and a million tons or more are removed every year, in several years of operation the whole operation will still be so small that you could walk the length of it in a few minutes. Mining experts who have looked at the problem consider the lunar mine so small scale that it will hardly keep one bulldozer occupied.

As long as we demand great quantities of elements which are not rare on the Moon, there will be no need for detailed assaying at the lunar surface. The average lunar soil (for example, the so-called "fines" brought back by Apollo 12) is about one third metals by weight, and almost a fifth silicon, useful for making solar cells to convert sunlight to electricity.[11] Oxygen is the most plentiful element on the lunar surface, and so will be an abundant and very useful "waste product" of the processing industry in space.

Television and personal report have shown us that men can work in space suits only slowly and inefficiently. If the lunar outpost is to carry out its tasks quickly and effectively we must plan the activity so that space suit operations are reduced to a minimum. The most time consuming task may be the assembly and checkout of the mass driver. Within a modest mass budget, a circular cylinder of aluminum large enough in diameter to serve as an assembly bay could be delivered to the Moon in sections, among the early payloads. In such a cylindrical tunnel, covered over with lunar soil for cosmic ray protection, the mass driver could be assembled and electrically tested.

By the time the cooks, the doctors, the communications experts and the other necessary service personnel are added, the lunar work force during the construction phase may total about fifty people. After construction is finished and the lunar outpost settles down to steady operation, the best estimates are that eight or ten people will be enough. On a typical work shift there may be one person monitoring the automated operation of the mass driver, while another controls a mining vehicle by television and radio. The two may be in the same room, at control consoles, and while the work goes on may be swapping stories and passing the coffee pot back and forth.

In most respects the lunar base will be the most remote and difficult to get to of all the locations where people will be working. It's unlikely to become a backwater, though. Scientists will visit the Moon both to do basic research and to carry out assays and surveying. Construction crews will visit each time the mass driver gets upgraded. As we now see it, the mass driver first located on the lunar surface will be capable of moving over a million tons of lunar material each year. Its power supply will be a lot heavier than the machine itself, though, so it makes sense to give it only a fraction of its final power initially, and add solar cell arrays as the industry in space expands.

When installed and operating on the Moon, the mass driver will launch its payloads at a slight downward angle. Their speed will be so great that they'll rise rather than fall, and after a free flight of a minute or so will be many kilometers away. There they will pass through a correction station, where their positions will be measured very accurately, and their speeds and angles will be corrected by the same electrostatic methods that are used in steering the beam of electrons in a television tube. The latest calculations show that after such steering the payloads will be able to hit a particular point in space with an error of only a few meters.

Climbing out against the pull of the Moon's gravity, the payloads will finally escape from it to free space at a relatively low speed. What's the best point to aim for? The best target seems to be the second Lagrange point, L2, out beyond the lunar farside. There a collector will be maintained in position, maneuvering to follow the slowly changing stream of lunar payloads as the trajectory changes over the period of a lunar month. When several thousand tons of material have accumulated at L2, they'll be ferried to L5 by a low thrust tug, and that tug may itself be driven by a small version of the lunar mass driver.

Newton's laws tell us that a machine which can accelerate and launch material with a high velocity can be used as a reaction engine, like any rocket. The mass driver, with its tons of steady force, will be quite effective for conveying large payloads in free space. Its performance, measured in terms of exhaust velocity, will be about that of the space shuttle's solid rocket motors.

The lunar machine isn't designed to be a rocket engine, and in the course of the intensive theoretical work on mass drivers I became interested in calculating the performance of a mass driver tailor-made to supply thrust, as a tugboat engine in space, driven by solar power. The numbers looked so attractive that in 1977 I included them in an article that is typical of our modern ways of approaching the space manufacturing problem.[12]

Let's be realistic about our plans of realizing the humanization of space. First of all, no one's likely to subsidize the construction of space habitats for their own sake no matter how attractive they may be. If they're built, it will be for the same reason that most new housing is built on Earth — there's an industry, or several industries, that need workers, and so a market exists for housing for the workers and their families.

If there is a need for products, in large tonnages, that will find their use in high orbit or beyond, we should search for the most efficient ways to set up the manufacturing and transport systems to build and relocate those products. How can we minimize the necessary investment? By using, as far as possible, the one vehicle system that is already under development: the space shuttle. During the decade of the shuttle's development, it's been planned for a traffic model ranging from 60 to 120 flights per year. If a particular orbiter must spend a long period in orbit, to carry out planned experiments, it can only be used in a smaller number of flights per year. To accommodate not only NASA's present (much reduced)

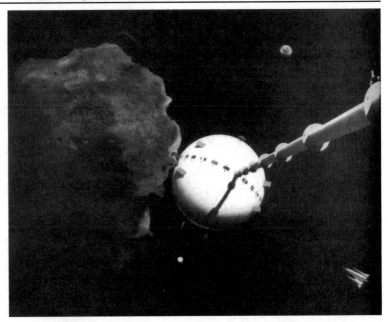

The mass driver, probably in a smaller version, could also be used as a solar powered rocket motor. Acting as a "tug boat" in space, it could use anything (such as processing slag and mine tailings) as its "propellant."

schedule of launches, but also a program of space manufacturing, some additions to the presently planned fleet of five orbiters may be needed. When the shuttle was first planned, it was thought of as a vehicle to lift components of a space station. More recently, as budgets have declined and the space station has shrunk to something more like a workbench in space, the shuttle traffic planning has been changed. Now the shuttle orbiters are thought of as doing double service, bringing experiments into space and remaining with them as temporary space stations. In terms of traffic efficiency, that's a bit like traveling to Europe in a 747 on a week's vacation, and then keeping the airplane on the ground the whole time in order to use it as a hotel. NASA has no choice, under present budget limitations, but if orbiters could be used literally as shuttles, bringing equipment into orbit and then returning as quickly as possible, a fleet of three or four more would be enough to double the number of flights, beyond the presently planned sixty a year.

In the article *The Low (Profile) Road to Space Manufacturing*, I outlined a way to attain a high level of production in space over a period of a few years, within a traffic model of about sixty flights per year of shuttles.[12] In the later years of that plan many of those flights would be of the shuttle derived HLV, the shuttle being retained mainly as a transport for people. Its cargo bay is about the same size and shape as that of a DC-9 aircraft, and if filled by a passenger compartment could carry — for the short flight into orbit — about the passenger load of a DC-9.

The *Low Profile* article built heavily on the results of a 1976 NASA study on space manufacturing. There, for the first time, we obtained the numerical data on the sizes and weights of processing plants in space, and on the number of people necessary for a manufacturing program with a certain tonnage of output per year. In 1977, in a much larger study, a group working under the direction of John Shettler of the General Motors Corporation followed up the *Low Profile* article by a much more detailed investigation, calculating the equipment payload and the passenger list for each flight. These are early steps in what is likely to become a continuing effort, as we search for the most cost effective ways of realizing space manufacturing. For that reason it doesn't make sense to list a great many of the published numerical results. Instead, I'll continue with the building blocks we now think of using.

All of the equipment we must locate on the lunar surface must first be hauled to lunar orbit, together with the rocket fuel needed to soft land the equipment. The shuttle can't do that job, and if we were to use a rocket powered tug, the shuttle would have to lift all of the fuel for the rocket. We plan a substantial cost saving by using a small mass driver, of very high performance, to carry out that inter-orbital transfer. The mass driver would be carried to low Earth orbit in several shuttle payloads, and would be assembled in orbit, and would from then on ferry equipment out to the vicinity of the Moon.

Where do we find, though, the reaction mass for the mass driver to throw out? It has to throw something, in order to develop thrust. The answer seems to be to use something that would otherwise be thrown away — the shuttle external tanks. The orbiter vehicle has engines (the SSME's) but no fuel tanks for them. When it rides into orbit it does so on the back of a much larger object, a big canister shaped like a fourth of July rocket. That canister contains hydrogen and oxygen tanks from which the SSME's draw their fuel, and when the shuttle is almost at orbital height and speed, that fuel is exhausted. The final tiny push into orbit is done by much smaller steering rockets carried by the orbiter, and at the moment of burnout the external tank suddenly becomes surplus, after a brief but glorious life of less than twenty minutes. It happens that the empty weight of the tank is actually greater than the whole shuttle payload, and it seems a shame to let that weight go to waste.

In the *Low Profile* plan, the tanks would be carried into orbit at a very small cost in shuttle payload. We would set up a storehouse of empty tanks in orbit. Some of these would be fitted out as living quarters, with each tank providing about twenty comfortable, private apartments for as many workers. In Shettler's plan, those modular apartment houses would turn up everywhere in the early days of space manufacturing — in low orbit, for the training and final screening of workers in the special world of zero gravity; in high orbit, for the work force tending the processing plants; at L2, for times when the mass receiver there may need repair; and on the lunar surface. As soon as lunar material became available in space, it would be used to shield the apartment modules from cosmic rays, and before that there would be minimal shielding, enough to protect against solar flares, made up of dehydrated foods stored for later use

One possible source for mass driver tug propellant is the Shuttle's large external tank. Normally expended in low-Earth orbit to burn up, they could easily be retained and used as "fuel" for a mass driver tug ride to higher orbits.

Most of the external tanks would end up as reaction mass, in pelletized or powdered form. In a typical (unmanned) ferry operation, several hundred tons of equipment, accumulated from shuttle payloads, would spiral up to lunar orbit, over a time of several months, with the expenditure of a somewhat greater amount of tankage mass, each small pellet leaving the mass driver tug engine with a speed much greater than that of a rocket exhaust. Dumping the equipment in lunar orbit, the mass driver would return in a much shorter time, arriving in low Earth orbit to pick up a load for another round trip.

In our present thinking there would be several stages in the setup of manufacturing in space, and if an insurmountable problem appeared at any stage the program could be terminated there. We don't really expect any such problems to appear, but it's far easier to arrange funding if there are well defined milestones and tests, each of which has to be passed before the final goal is reached.

The first stage is the setup of the lunar mass driver and the start of the transporting of lunar materials into space. That seems to require only about two years' worth of shuttle flights. Once that milestone is passed, we'll be able to bring into high orbit about ten times the amount of material that the shuttle can lift. Already from that point on there'll be plenty of mass for shielding, and plenty of "fuel" for mass driver reaction engines.

The second stage is the beginning of chemical processing of lunar materials into pure metals, glass and oxygen. That takes about another year's worth of shuttle flights, to lift the processing equipment, solar power arrays to run it, and other essentials. When that stage is reached the number of workers in space will be something like one or two hundred.

Now comes another application of the "bootstrap" method. The most complicated and sophisticated pieces of equipment needed in space — things like mass drivers and chemical processing plants — turn out to be rather light. It makes sense to build them and test them on the Earth, and lift them to orbit with the shuttle. The heaviest pieces of equipment needed in space seem to be solar cell arrays needed to power both the lunar mass driver and the processing plants. The first pilot plant in space will already be turning out each year several thousand tons of metals, silicon and oxygen. We plan to use all three to bootstrap our way to a much higher level of productivity.

The metals and silicon will go into solar cell arrays. Those we will use to upgrade the tonnage per year that the lunar mass driver sends out, and to equip new duplicates of the original space processing plant. The oxygen will be used in several ways: as the heaviest part of the fuel burned by the rocket tugs and landers; as the heaviest part of the water that will be needed by the work force in space; and as an ideal reaction mass for the increasing traffic of mass driver tugs hauling freight in space.

It seems that by this kind of cost saving approach we can build up to a level of processing a million tons or more of lunar material each year, over a period of seven or eight years, without ever exceeding the lift capabilities of the shuttle. How about the economics? In our traffic model we'll be paying about $1 billion in shuttle launch costs each year, over a seven year period. At the end of that time, though, we'll be producing about a third of a million tons every year of finished products, and relocating them either in geosynchronous orbit or wherever else in nearby space they will be used. A good cautious estimate would be to assign those

For both passengers and freight, development of advanced rockets is needed. Single-stage-to-orbit, completely reusable rockets capable of lifting higher payloads than the Shuttle, at a much lower price per pound are possible with 21st century technology.

finished products a value of around a hundred dollars a kilogram. The lift costs alone, to bring a kilogram into high orbit, are in that general range even for very advanced, totally re-usable rocket concepts many times larger and many years later in time than the shuttle. With those numbers, the manufacturing facilities in space will be producing $30 billion every year in value. A bargain indeed.

How soon could it all happen? Both in 1976 and in 1977 the studies independent of NASA, but supported by that agency and closely cooperating with it, worked out program plans based on slow and fast rates of decision making. It seems to be agreed generally that there are far greater uncertainties in the political decision making process than in the technical areas. With rapid decisions, both the '76 and '77 studies agreed that the first liftoff of equipment destined for space manufacturing could occur as early as 1985, and that the first substantial payback in the form of products made in high orbit could occur as soon as 1991. On that time scale, by the mid-1990's the construction of Island One, as a more comfortable, long term habitat for manufacturing workers and their families, could be done almost as an aside, with the diversion of only a few percent of the manufacturing productivity that would exist in space at that time. There's no answer to the question "What's the longest it might take," of course, except "Never." The more leisurely program plans put Island One around the year 2010. To those of us who feel that space manufacturing offers great potential for human benefit, such a delay seems nearly criminal, but on the time scale of human existence a mere fifteen years is hardly the blink of an eye.

# *Chapter 9*
# *First Tasks For Island One*

As the first new world in space takes shape, over a period of several years, surely the moment of "sealing" will be planned for and celebrated. Oxygen long stored in liquid form will be allowed to enter the sphere, and pressure throughout the living and agricultural areas will slowly build toward its final value. Many of the construction workers may move their activities to the new villages at that time, and enjoy the luxury of roomy surroundings as they complete the apartments and other buildings.

As they work, a small electric motor the size of an automobile engine will apply its power to rotating the habitat, until finally, after several months, the gravity at the equator will reach Earth-normal. By then, the soft green of growing plants will have turned the valley into something very like a small patch of farmland in springtime.

With the greening of Island One, and the harvesting of its first crops, the long term residents will come. And for many in the construction work force a time of decision will arrive — a time when the choice must be made, to return to Earth or to stay on to help lead the growth of the new communities in the first permanent human world beyond Earth. Many will choose to return to our planet, to spend and enjoy accumulated earnings. Some though, will probably feel that nothing here can offer them the excitement and the challenge of construction at L5. If human nature and history are guides, some people will make the first choice, will visit for a time on the Earth, and then will be outward bound again to rejoin their friends who may never have left.

Island One, though modest in size, may be an attractive place in which to live and work. Certainly there will be few communities whose citizens have so many talents and so much determination. Whatever the attractions of Island One, if it is to take its place as part of the complete human world, it will have to produce, more effectively and efficiently than can be done in any other way, products which are needed urgently by the rest of the family of humankind.

Island One will have a unique economic advantage for just one class of products: those whose end use is in free space or in high orbit above Earth. When we attempt to build such products and launch them from our planet, we must pay heavily in energy. Here on Earth we are the "gravitationally disadvantaged," located as we are at the foot of a gravitational mountain some 4,000 miles high.

For any product whose use is to be in or near free space, high above Earth, production at L5 will save the lift cost — many dollars for every pound produced. A worker at L5, producing at a similar rate to our heavy industry on Earth (more than twenty tons per year) will be turning out a value of several million dollars per year beyond the intrinsic value of the goods, simply because of the saving in lift costs from the Earth. The "Swiss banker" approach to estimating the value of Island One is the most conservative we can imagine — value the goods produced by taking the lowest possible lift costs for a competing industry which must lift its products from the Earth, and don't assume any extra productivity in space even though we know that zero gravity and automation are almost certain to favor high production. When that is done, and only half the population of Island One is assumed to be engaged in factory work, the products of Island One still come out to be of great value: many billions of dollars each year, enough to pay back the investment in a very few years.

In the very long term, perhaps material products or raw materials from space can be returned usefully and economically to the surface of Earth. That seems to me a rather unlikely prospect — at least for some time — because if L5's industries begin manufacturing material products to be used here on Earth, they will give up their single greatest advantage — their location at the top of the 4,000 mile gravitational mountain at

Construction workers building Island One will most likely be the first people to make the decision to become future residents of a space habitat.

whose feet we now stand. Similarly, I see no great advantage in L5 for the zero gravity processing of very lightweight, high value products. For those, it makes more sense to lift the raw materials from Earth to low orbit, by way of the shuttle, and then to return them to Earth the same way when they have been formed in zero gravity. Others have estimated the total market for products of that kind — vaccines, single crystals and other exotica — and have concluded that even twenty five years from now that total market will be so small as to require only a few shuttle flights per year for its satisfaction.

Before considering the major industries for L5, we should ask first whether some of the benefits from Island One's construction may appear during the time of its building. There may well be such benefits, and my guess is that they will be mainly scientific. Once the lunar outpost and the L5 construction station are established, with all of their facilities for supporting people, for transport and for communications, they will also be locations ideal for other work than that of producing Island One. At their locations scientific research can be carried on at far lower cost than by the exquisitely complex, delicate pieces of "orbital jewelry" which we now have to launch for completely unattended automatic operation. I expect that in addition to the eight or ten people of the mining and transporter servicing outpost community on the Moon, at any given time there may well be several geologists and other scientists in long term residence. These people could spend half or more of their working time on such practical tasks as bore sampling the lunar surface, assaying minerals and planning the optimum locations for materials gathering. The rest of their time they could spend on pure research as opposed to applied. If our experience on Earth is any guide, these two activities would be separated only by an indistinct boundary, and would reinforce each other — knowledge gained in one area often finding its greatest use in the other. At L5, by the time the work force has built up to several thousand people, even before Island One is built there might well be fifty or one hundred whose tasks would be wholly or mainly scientific. Some could be maintaining and gathering data from large space telescopes, located just far enough from the station so as not to be occulted by its busy transport craft, but close enough to be reached in a few minutes travel.

Of the aluminum and other metals being produced at the station it would be surprising if some few percent were not allocated to scientific purposes. The first of these might well be the construction of large optical and radio telescopes. I don't think it likely that such scientific efforts, greatly as they would benefit by "tagging along" on the main construction activity, would ever enjoy budgets large enough to pay back a large fraction of Island One's construction cost, but their scientific objectives could be reached at far lower cost because of the existence of Island One's construction station.

For the scientists themselves, the presence of the L5 construction station would certainly constitute a great boon. Typical scientific programs for space research, even those which involve unmanned satellites, cost several tens or hundreds of millions of dollars. In contrast, the "exchange" cost of sending a scientist to L5 could be as little as a few hundred thousand. One gets a "ticket cost" in that range by taking the existing space shuttle as a passenger carrier, with the published NASA figures for cost per flight, and assuming that the transfer from low orbit to L5 is made by a conventional rocket powered tug whose heaviest fuel component, liquid oxygen, is obtained as a waste product from the processing of lunar soils at L5.

Years later, when more efficient vehicles are developed, we can expect that the costs for passage from the Earth to L5 will be reduced, ultimately to only a few thousand dollars.

The most recent studies agree that in the early days of the buildup of production capacity at L5 it will be more economical to bring food from Earth, rather than to attempt to set up agriculture in space. By the time the work force reaches several thousand people, though, supplying their food from the Earth will begin to strain the capacities of the shuttle derived HLV at its normal "traffic model" flight frequency. In detailed studies the trade-off between resupply and space agriculture has already been calculated, and it seems fairly certain that by the time of Island One the people who are living in space will be growing most of their own food. Very similar arguments come up when the planners set out the tours of duty for the early construction workers. It seems that we are likely to begin with stay times in space of a few months to a year, and then will gradually extend to stays of two or three years, family members accompanying each worker. Clearly the balance between exchange time, the degree of luxury of the construction station, and the salaries paid to the construction crew will have to be made with care after considerably more study.

The problems we now face here on the surface of Earth due to the rapid exhaustion of conventional fuels were described in the first chapters. There are natural sources of energy which we do not now fully exploit, and which could be of benefit to us in extending the fuel reserves that now remain. These include geothermal energy, hydroelectric power, the winds, the tides and solar power. All of these "exotic" sources of energy have serious limitations. Either they are undependable, or the capital cost of using them is too high, or (as is the case particularly with hydroelectric power) their further exploitation could only be accomplished at a very serious ecological and environmental cost.

Two sources of power for the future are now under intensive study: nuclear fission, particularly in the form of liquid metal fast breeder reactors; and hydrogen fusion, by magnetic containment of a plasma or by laser implosion of small deuterium-tritium pellets. It would be rash to attempt to guess the probability that one or both of these methods will turn out to be economically viable. Fast breeder reactors would have a decided environmental impact, and would also affect the political tensions of the world in ways on which we can only speculate. Rather than guess how successful, how acceptable, or how economical one or both of these methods might turn out to be, I will say only that both are high technology options on which research is now very active. At present, at least $700 million of government money is being spent each year on nuclear energy research in this country alone.[1] Of that amount, most goes into fission research, the remainder to fusion. One of the difficulties with the breeder reactor option is that the "doubling time" for converting non-fissionable elements into usable nuclear fuel is estimated to be at least ten or twelve years, while the world need for new energy resources is doubling in a much shorter time. As for nuclear fusion, most responsible scientists working on it hesitate to claim that it might be economical, even if it can be made to work, in less than about thirty five years. It does not seem to me very likely (and here I express what is necessarily only my own opinion) that either will be able to reduce significantly the cost of electric power. The proponents of the two schemes usually argue at most that one day they might be at a par economically with current fossil fuel plants.[2]

Perhaps surprisingly, it appears that Island One may be in a uniquely favorable situation to provide for us, here on the surface of the Earth, an alternative energy source which might be simpler, cheaper, and more acceptable environmentally than the first two alternatives. The space manufacturing facility could do so by building Satellite Solar Power Stations (SSPS). Satellite Solar Power is a concept that originated in the 1960's, and whose most active champion has been Dr. Peter Glaser, of the Arthur D. Little Company in Cambridge, Massachusetts.[3] The plan consists of locating in geosynchronous orbit, above a fixed point on Earth's surface, a large solar power station. At the station solar electric power would be converted to microwave energy, which would then be directed in a narrow beam to a fixed antenna on the ground.

Solar power stations will be the incentive for both investors and governments to participate in the earliest space colonies. The colony residents will realize a source of income by constructing and operating these power stations.

At first glance this scheme appears impractical. Without calculation, most engineers would assume that the inefficiencies of conversion, transmission and reconversion would be so low that no such power station could be economically viable. Curiously, the transmission problem seems to be solvable. Research on high power microwave transmission has demonstrated experimentally that power can be transmitted at an overall efficiency of at least 55 percent.[4,5] The target figure for economic viability is not much higher than that, so with moderate development one would expect the target to be attained. The environmental problems of microwave power transmission will have to be studied carefully, but so far they seem to be much less severe than those of radioactive waste generation from fission or fusion nuclear plants. The microwave beam would arrive at Earth with a beam width of about seven kilometers. Its intensity would be modest, less than half that of sunlight. In contrast to sunlight, though, it would be there all the time, even at night or in clouds or rain, and it would be in a form ready for conversion to DC current with a loss of only 10 percent. The antenna region on Earth would be fenced, and outside the fence the intensity of microwave radiation would be no higher than outside a microwave oven with the door closed. One or two kilometers farther away it would be far lower still. Although the beam would be no "death ray," studies would have to be made to be certain that it would have no long term effect on birds flying through it frequently or nesting in the antenna, and that it would not damage the communications radios of any aircraft straying into it.

Satellite solar power would have significant advantages over its possible competitors, beside the fundamental one of generating no radioactive wastes. Because the conversion of microwave energy to direct current could be done with such high efficiency, only a very small fraction of the total power would be released as waste heat into the biosphere from such an installation. In contrast, generator stations using fossil or nuclear fuels deposit as waste heat in the biosphere about one and a half times as much energy as they put into the power grid.

The market for new power stations during the time when Island One could become productive has been estimated by a number of task groups. For the United States alone, even assuming energy conservation, there will be a need for 65,000 megawatts per year of new generator capacity in the year 1990, and substantially more than that each year a decade later. For scale, the largest single power plant that one normally sees in driving the roads of America is about 1,000 megawatts. The cost of new power plants fueled by coal is roughly half a million dollars per megawatt, and nuclear plants are considerably more expensive. Consequently, the market for new power plants in the United States alone, assuming prices for coal fired generators, will be about $33 billion in the year 1990. A satellite solar power station requires no fuel, so its market value may be similar to that of a hydroelectric station of similar size. One of the largest and newest of the hydroelectric installations in the Western world is Quebec Hydro at Churchill Falls in Canada. Its price per kilowatt is about three times that of a coal fired plant, but because it requires no fuel it can supply electricity at a very low rate. On that basis the market for new satellite power stations in the United States at the end of this century turns out to be well over $100 billion per year.

If we include, as we properly should, the additional market represented by the remainder of the industrialized world, and provide for the needs of the nations now struggling to industrialize, the requirement becomes much larger still.

For any power source requiring a large development investment, the potential for long term growth is important. The SSPS concept appears to fare well on that score also. In the extreme case (certainly not realizable in practice) that SSPS power were to become the sole source of electric energy in the United States in the year 2000, the land area necessary for the SSPS antennas would still be only 0.2 percent of that of the continental United States — that is, about one fifth of the area already devoted to roads. Unlike the roads, SSPS antennas could be located in remote areas where they would not be visually obtrusive. They would be almost fully transparent to sunlight, and would block out microwaves from the land below them, so the areas below them should be usable as protected grazing land.

By contrast, if solar cells at Earth's surface were to be used to supply all our electric power, we would have to cover about forty times as much area, or 8 percent of the continental United States, with opaque solar arrays. The reason is that solar cell electric conversion efficiencies are about 16 percent (instead of 80 percent) and that the average over a year of solar energy intensity in the United States is only an eighth as much as in space.

If satellite solar power is an alternative as attractive as this discussion indicates, the question is, why is it not being supported and pushed in a vigorous way? The answer can be summarized in one phrase: lift costs.

I have discussed the present and the hoped-for figures for lift costs to L5 from the surface of Earth, based on present rocket vehicles and on those which could be developed at low cost with existing engines. Estimates by NASA center on about two hundred dollars per kilogram, for the shuttle derived HLV. If we don't "go for the Moon" and bring out lunar soils as reaction mass for mass drivers, we're forced to bring up from the Earth all the fuels needed for the lift from low orbit to geosynchronous. In that case the lift cost to the final location of a satellite power station will be several times higher than to low orbit. (The velocity change needed in order to bring a payload from Earth to geosynchronous orbit is about the same as to L5, so lift costs to either destination will be rather similar also.)

Large power plants could be built in either of two ways: as turbogenerator stations, like present day generators on Earth, or as arrays of solar cells, converting light directly to electricity.

For use in a power satellite, the most suitable variety of turbogenerator is a "closed cycle Brayton" system, in which gaseous helium recirculates endlessly between a heater, a turbine, and a radiator.[6] Such systems are rather light and compact as turbines go. Fortunately one such machine has been installed at Oberhausen, in West Germany, and has been working since early 1976.[7] It is heavily instrumented, and will provide plenty of operating experience on which future performance estimates can be based. Studies by the Boeing Aircraft Company, under NASA sponsorship, indicate that a power satellite based on a turbine of the Oberhausen type (that is, right-now technology) would have a mass of about ten tons per megawatt of output power. There is hope, but so far only a hope, that by pushing temperatures higher and using more exotic materials at critical locations that figure can be reduced.

We can assess the state of the silicon solar cell art by the fact that such photovoltaic power supplies in operational satellites of the past decade have weighed about ten times as much as an Oberhausen-type turbogenerator.[8] For the Solar Electric Propulsion System space probe scheduled to fly in the mid-1980's,

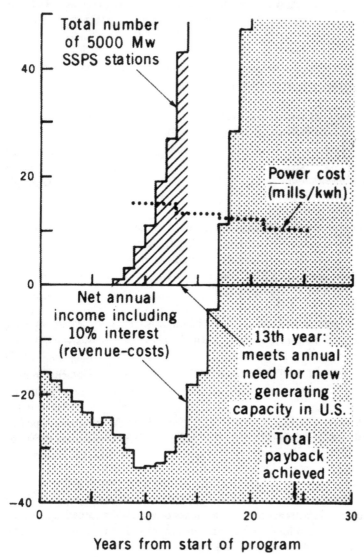

Supplying Earth with energy via solar power satellites, being built at a reasonable constant rate, would show energy costs on Earth beginning to drop within a few years of program start. The program could be meeting all of the US's power needs in little more than a dozen years, and from an investment viewpoint, the project would break even in less than 25 years.

NASA is hoping to bring the mass of solar cells down close to the Oberhausen generator figure for tons per megawatt.[9]

If we take the Oberhausen figure for performance, together with a transmission efficiency of around two thirds, and lift costs to geosynchronous orbit characteristic of the shuttle derived HLV and a rocket tug, we find a transport cost of $13 million per megawatt of installed capacity, lifted to geosynchronous orbit. That is many times larger than for the most expensive power plant now thought of for Earth.

The proponents of satellite power recognize this fact, have represented it accurately in discussions of the topic, and have sought to circumvent the problem by encouraging the vigorous development of lightweight silicon solar cells. Solid state research moves rapidly, and it may be that ultimately great reductions in the weight of solar cells will be brought about. Even the most optimistic estimate does not suggest, though, that they could be reduced in mass by a factor large enough to make the Earth launched SSPS concept viable without two more developments: first, while solar cell mass per megawatt of power is being reduced manyfold, their cost must go down by an even bigger factor; and in addition to these improvements, lift costs to geosynchronous orbit must come down to a tenth or so of what we could get with the shuttle derived HLV. To achieve that, it would be necessary to develop advanced space transport systems for which an investment of several tens of billions of dollars and many years of time would be required.

In giving these figures, it is not my intention to deny the possibility that all of these improvement factors could be achieved; I simply do not know. Nor is it my intent to discourage or delay the development of a prototype SSPS — any new technology requires a learning process, and if the basic SSPS concept is to become usable, that learning process must go on. Rather, my purpose is to explore an alternative method of the quantity production of economically competitive SSPS units.

Given the existence of Island One, it could produce a satellite solar power station, from lunar surface materials, within the technology limits of the present day. It could simply build large turbogenerators. A complete power station built around a Brayton cycle turbine would start with solar mirrors, concentrating sunlight onto boiler tubes. Helium brought to a high temperature in these tubes would pass through the turbine, then to a radiator, and be recirculated for another passage. The turbine would drive an electric generator of the conventional sort now found in Earthbound power stations.

If this design is followed, a station built up of several large turbogenerators will be connected to a disc shaped transmitter antenna. The conversion from low frequency to microwave power can be made by a large number of small tubes, each like those which power microwave ovens. Operating in the vacuum of space, these tubes will have no need for glass envelopes.

If we take present day figures for the masses required, a station able to supply 5,000 megawatts to a national power grid on Earth's surface will total some 80,000 tons. It can be assembled and tested as a single unit, in zero gravity just outside Island One. The work force assembling it will be able to return after each day's work to the comfortable Earth-like surroundings of their habitat.

Studies by the NASA Johnson Space Center, based on projections of technology rather than on the right-now situation, are about twice as optimistic as these figures on mass per megawatt. If they are correct, Island One could turn out about twice the value per year that I've estimated.

The space manufacturing site will be some distance from geosynchronous orbit. The costs of transport in space, however, are measured not in distance but in velocity interval. In those units even L5, the most distant of the possible sites, is closer to geosynchronous orbit than to the lunar surface. To move so large a mass over the required distance will require a mass driver, and it could be identical to the one already in use on the Moon. The steady four ton force produced by that transporter will be quite enough, over a period of months, to move the power station into its position high above a fixed point on Earth. The electric power input to the mass driver will come from the station itself. The necessary reaction mass to carry out the transfer can be industrial slag, pulverized rock dust, or liquid oxygen, all of which will be available at L5. Return of the mass driver to L5 for reuse can be made with the help of a small solar power plant. A power plant only about a thousandth the size of the SSPS itself will be quite enough to return the mass driver to L5 for re-use in a month, so a mass driver used as a tugboat for SSPS barges can make several round trips every year.

It was pointed out to me by Mark Hopkins, a young economist from Harvard, that the economics of SSPS construction at L5 requires a fresh viewpoint. In that construction almost no materials or energy from Earth will be needed. Island One, when it is established and operating, will be self-sustaining, and its residents will be paid mainly in goods and services produced at the space community.

The economic input to a combined space community / SSPS program will be the sum of the development and construction costs for Island One, the cost of lifting the material needed from Earth for subsequent communities and for those SSPS components which cannot be made at L5 economically, a payment on Earth to the credit of each person living at L5, representing that portion of salaries convertible to goods and services on the Earth (for subsequent use on trips or, if desired, on retirement) and a carrying charge of interest paid on the outstanding balance in every year of the program.

If Island One and its sister colonies become the main source for new generator capacity to supply electricity for the Earth, the question of legal ownership of the SSPS plants ties in to the economics. Geosynchronous orbit is far below L5, and I suspect that any Earth nation using SSPS power will want clear-cut legal ownership of the power generating facility once construction is finished. From then on that nation will control the power station and any maintenance operations on it, and will keep the SSPS fixed above a certain point on its own territory, where an antenna is located.

If Island One were to be independent of Earth, it would also be to the economic advantage of the workers in space to sell completed power stations rather than electric power. In that manner they could get a quicker return. From the viewpoint of the nation, consortium of nations, or consortium of utilities which might provide the investment capital to build Island One, it's more cautious though to assume that the only

Numerous technical and financial studies have been made, both by governments and private industry, that validate the solar power station and Island One proposals. Models of the Bernal sphere (Island One) have been officially displayed for many years and have undergone no fundamental changes.

economic payback will occur from the sale of power at the transmission lines on Earth. For many reasons, among them legally binding treaties which have already been signed by several nations, it seems wisest to assume that initially Island One will be tied to the Earth governmentally.

By now the economics of SSPS construction at space manufacturing facilities has been discussed in a technical article 10 in testimony before Congressman Donald Fuqua's subcommittee of the U.S. House of Rep-resentatives,[11] Senator Wendell Ford's subcommittee of the U.S. Senate,[12] and in testimony before the Energy Commission of the state of California and the Energy Research and Development Administration of the federal government. Those economic projections have always been on the cautious side, assuming high lift costs for the space manufacturing equipment, large mass for the SSPS plants, relatively low productivity in space, high interest rates on investment, and low electric rates for SSPS power supplied to the Earth. Yet all the projections confirm that SSPS plants built at a space manufacturing facility out of non-terrestrial materials should be able to undersell electricity produced by any alternative source here on Earth.

The most recent studies, based on the *Low Profile* approach to space manufacturing, appear even more attractive, because they indicate useful production starting well before Island One is built, at a time when the total investment is much less than the $100 billion estimate originally made for Island One.

By now our planning group benefits from the advice of senior executives in the electric utilities and investment communities. From them we have learned a good many realities that help us in guiding our research. For one thing, it seems almost certain that we cannot expect private capital to invest in space manufacturing until the risks have been reduced almost to zero. Government funding, possibly by a consortium of several governments, will have to carry the program at least until a pilot SSPS, not necessarily made from lunar materials, has supplied energy to the Earth. At the same time, we will have had to demonstrate that we can bring out lunar materials and process them in space, to get the same elements used in building the SSPS. Above all, the economic studies made at that time will have to show that SSPS power can undersell all competition. Once that happens, though, it appears that private capital in large quantities should be available for the expansion of the program to full capacity.

As of the late 1970's the lowest price electric power in the U.S. costs around two cents per kilowatt-hour at the power plant. Our goal is to undersell that price, whether or not the prevailing rate goes up in later years.

As the possibility of construction of Island One is examined with greater care and in more detail, both the engineering and the economics can be studied in far greater detail than they have been. The most significant point about this discussion, though, is that already it can be carried out at the level of engineering and economics. There is no dependence on any basically new physics, nor on any great extrapolation of engineering practice beyond what is customary today.

One of the graphs prepared for the evaluation of space manufactured SSPS plants was given in congressional testimony (Appendix II). According to that graph, rather quickly, within thirteen years from the initiation of heavy investment in Island One, the rate of construction of new generator capacity in space could exceed the annual growth needs of the United States. Not long afterward, the total energy so far supplied from space could exceed the total stored in the Alaska North Slope in the form of oil.[13] The contrast is glaring: in the case of Alaska, at that time there would be little remaining from the Alaska pipeline (except perhaps for oil slicks on the water and some discontented elk) while satellite space power could still supply clean electrical energy to the Earth for another five billion years — the estimated life of the Sun.

For an enterprise demanding investment capital at the start, and yielding profits at a later time, economists calculate what is called the "benefit / cost ratio." Taking account both of interest charges and of inflation, the benefit / cost ratio summarizes whether a possible investment is worthwhile or not. Even without the cost savings of the *Low Profile* approach, the benefit / cost ratio for space manufacturing is much above one, indicating that in spite of high interest rates and low power charges, the Island One program would be a paying proposition. To get that favorable result, it appears that exponential growth of the manufacturing capability in space is very important; a slower, linear growth doesn't pay off fast enough to make up the interest charges on investment.

When the amortization of power plants is complete, the cost of power generated in this way should go down, because the satellite stations should require little maintenance and will be using free energy — given by an efficient, clean thermonuclear reactor which has been located for us at a comfortable distance of 150 million kilometers.

If this development comes to pass, we will find ourselves here on Earth with a clean energy source, and we will further improve our environment by saving, each year, over a billion tons of fossil fuels, now lost to heat and smoke in driving our electric generators. Given a worldwide market which may be several hundred billion dollars by the year 2000, probably the industries at L5 will grow rapidly in numbers and size, to satisfy so urgent a demand.

If satellite power stations are built at L5 rather than on Earth, there will be important environmental consequences. For every SSPS that would have to be lifted from the Earth if built here, many times as much weight would have to be dumped into the atmosphere in the form of rocket fuel exhausts, to lift the SSPS components. The total quantities run in the range of hundreds of millions of tons per year, if SSPS power becomes dominant in the world economy. No one knows what the environmental effects of those exhausts would be, but it seems sure that writing the "environmental impact statement" for such a program would be no easy task. By contrast, the establishment of space manufacturing requires only about a hundredth as much lift tonnage from the Earth, and is within the existing space shuttle traffic model, whose impact has been carefully studied and shown to be safe.

A major open question, of course, is what fraction of the mass of an SSPS power plant couldn't be obtained from non-terrestrial materials, and so would have to be lifted from the Earth anyway. If we were using asteroidal materials, we could be sure of having in quantity all the elements we have on the Earth. The Moon, though, is poor in hydrogen, nitrogen, carbon, and some heavy metals. Fortunately NASA has now begun to study that question, and over the next years we may hope to see designs for satellite power stations optimized for the use of lunar rather than terrestrial materials.

So we're beginning to perceive a possible branch in the development of satellite power. The best game plan seems to be to keep the options open — build small pilot plants, improve solar cells, and meanwhile push the early research and development of mass drivers and lunar soil processors. After a few years of research, when the numbers are clearer, there will then be a point where a rational decision can be made, either to develop the huge, very advanced lift rockets needed for a satellite power system built on the Earth, or to put a similar amount of money into developing the "non-terrestrial alternative."

If the efficiency of industry in space improves to the extent predicted by some students of the subject, the cost of satellite solar electric power delivered on Earth could drop to much less than one cent per kilowatt-hour at the antenna on Earth. If that occurs (I am not yet willing to claim that it will, because research has

Using materials from space for solar power station construction and lifting from Earth only those materials not readily available in near-Earth space would require a launch rate that is within the existing space shuttle traffic model. Lifting all materials from Earth would have an incalculable negative environmental impact.

not yet been done in sufficient detail) it would have profound consequences for international politics. With low enough electricity rates, it would be possible to synthesize clean artificial fuels, which could compete economically with gasoline and render this and other participating nations independent of oil imports.

It would be well within the capabilities of Island One to produce a large optical telescope made up of many individual mirrors. Great resolution could be achieved by locating the individual elements of that telescope in a precise array stretching over a considerable distance, instead of combining all the mirrors together.[14] In designing such a system it is natural to consider linking the mirror elements by a mechanical structure, but that might be the worst possible thing to do. A mechanical link would expand and contract with temperature changes, altering the mirror spacing. It might be preferable to take advantage of a zero gravity location by building a large number, perhaps several thousand, of individual glass mirrors, each a meter in diameter, and providing each with a small locator module, equipped with station keeping gas jets. The heavy parts of such an array might be made at the space community, while the light, complicated, labor intensive parts would be brought up from Earth.

If the elements were linked only by light beams, their spacing could be established by the unvarying number of wavelengths of light between each pair. That nonphysical linkage, computer controlled, would have the further advantage that the mirrors could be programmed to separate and reform, like dancers in a slow motion ballet, according to the needs of a particular astronomical experiment. If located in a cross shaped array, with individual elements spaced ten meters apart, a telescope of that kind would have the theoretical capability of resolving something as small as a changing weather system, one thousand kilometers on a side — on the planet of a star ten light years away!

Once Island One is in full operation, almost surely the scientists will argue strongly to have part of its production capacity put into ship construction. Even a percent or so of L5's output of aluminum, magnesium, titanium, iron and other useful metals will be enough, over a period of two or three years, to build a large research vessel which could be in many respects a space borne equivalent of Darwin's ship the *Beagle*. Equipped with an engine which might be a slimmer, longer version of the lunar mass driver, this research spacecraft could voyage to an asteroid, using as reaction mass crushed rock dust. The *Beagle II* might have a crew much larger than the original H.M.S. *Beagle's* fifty, and they would form a small, self-sufficient laboratory village. The "launching" of their craft would require none of the flame and thunder that accompanies a launch from Earth. Instead, floating at rest in space at the entry dock of Island One, the vessel would seal its entry port and quietly cast off. Let us imagine the voyage as though its details were certain:

When power is fed to the engine, the ship will begin to move almost imperceptibly, hardly an arm's length in the first half minute. But a day later it will be only a small dot of light in a telescope, and after a month it will be ten times the distance of the Moon.

When its crew, many months later, performs a rendezvous with a small asteroid, the scientists aboard will take as much time as they like to study the planetoid in great detail, measuring its mineral content, assaying its resources of carbon, nitrogen and hydrogen, and collecting tons of samples. Much of their work will have direct "applied geology" applications to the later use of asteroidal material for construction — that is probably what will pay for the trip. Other work, carried on a fraction of the time, may seem then to be without direct application; the view from later on, a few years afterward, may be quite different.

While the scientists are at work, the engineers will use their on-board machinery to mine and collect several thousand tons of rock and dust as reaction mass — fuel for the next leg of their voyage. When the voyagers cast off and feed solar power to their engine to return or go farther afield, much of the scientific information collected will already have been analyzed and radioed back to L5 and to Earth. During the long months of the voyage that follows, samples will be analyzed in the ship's laboratories, information digested, and scientific papers written. Submitted from deep space by radio transmission, these papers may bear such identifiers as: *Carbon 12 / Carbon 13 Analysis for Asteroid 2655; by —, Beagle II Research Laboratory, en route to Ceres.*

As the voyage proceeds, a small on-board rock crushing plant will run continuously, providing rock dust as reaction mass for the engine. Unless the crew finds it too confining to be isolated in a small traveling village, a vessel of this sort could cruise among the asteroids for years. Surely families will travel together, and children will go to school on the *Beagle II*, sharing work and relaxation with their parents. Later, in the days of Islands Two or Three, much larger ships can be built, carrying with them to the outer regions of the solar system whole research institutes and sections of universities.

It seems likely that Island One will become a favored place for scientific sabbaticals from Earth. Especially for young scientists, not yet concerned with marriage and family, the opportunities for research at the space habitat in radio and optical astronomy will be unexcelled. A cycle may well be set up in which a scientist will arrive for a year of intensive data taking, then exchange his place with a new arrival, while he returns to Earth to analyze his data and write his conclusions in article form for the scientific literature.

For research in radio astronomy, most antennas are arranged in geometrical patterns, like crosses or circles. One special type of antenna, though, might be formed as a huge parabolic dish. I confess to some misgivings about the use to which this great mirror would most likely be put, yet I cannot deny that Island One would be the ideal place for its construction and use. This antenna would be used for a project known as *Cyclops* — the big eye; a search for extraterrestrial civilizations.

For more than fifteen years there has been interest in the possibility that there may be other intelligent civilizations in our galaxy, which could be members of what some people have called the "Galactic Network."[15] It's difficult to say, on the basis of any theory, what the chances are that such civilizations do now exist. The idea that we, as (to some degree) intelligent life are unique is of course absurd — the more we learn about the origins of life, the more we realize that the conditions under which life first began on Earth must have been duplicated many times over in other parts of the galaxy. The crucial unknown quantity, though, is outside the natural sciences entirely: it is the lifetime of a communicating civilization.

Our galaxy is disc shaped and has a volume of a thousand billion light years. Many of the individual stars within it may live in a stable manner for several billion years. In the modern view, perhaps one in ten of the hundred billion stars of our galaxy may have planets, and so be "likely" places near which life may originate. In 1959 Phillip Morrison and Giuseppe Cocconi speculated on the possibility of searching for extraterrestrial life with the sensitive receivers used in radio astronomy.[16] Soon afterward, Frank Drake carried out the first search intended specifically to look for such intelligently directed signals. His *Project Ozma* was capable only of examining a few nearby stars, and found only natural signals.[17]

Those scientists most interested in the search for intelligent extraterrestrial life recognized some time ago the importance of two vital numbers: the odds that such life will develop on a planet of a "likely" star and the length of time that a civilization will be actively engaged in radio communication. The importance of these two numbers can be illustrated by examples: If life in the galaxy is very abundant, then perhaps as many as one in ten stars with planets become the nurseries of new civilizations at some time in their evolutionary history. If so, there may be as many as 100,000 stars within 1,000 light years of our Sun, each of which becomes at some time the birthplace of a civilization. What is the chance that we, searching all the one million planet-bearing stars within that great sphere, will find at least one which is beaming signals toward us? That depends very much on the second critical number — the duration of communication. Even if the average civilization remains actively engaged in communication for 100,000 years, and even if it devotes to that purpose an effort sufficient to beam signals continuously toward every likely star within its own 1,000 light year "sphere of interest," the chances are only about even that we are on the scene at the right time to receive an intelligent signal. The reason is that for any given civilization the period of communication corresponds, in our example, to a brief moment of time which occupies only one part in 100,000 of the whole evolutionary history of its star.[18]

Space colonies like Island One are not only a necessity to accommodate our overcrowded, depleted Earth, but may also be the "next step" in evolution.

The uncertainties connected with such numbers are so great as to leave open two extreme possibilities. First is that communicating life is sparse, that the duration of communication is rather short on the galactic time scale (by short, I mean 100,000 years or less) and that we are, therefore, at this moment alone within a 1,000 light year distance, or even alone in the galaxy.

The other extreme case, still open as a possibility, is that the galaxy teems with communicating life, that the duration of the "attention span" of civilizations is many billions of years, and that consequently as soon as we put our ears to the ground we will hear the beat of distant drums.

With so much room for the imagination I find it irresistible to add my speculations to those of so many others who have written on this topic. My guess, and it can be no more than that, goes this way:

First, I think that soon after a civilization reaches our own modest level of technological competence it becomes unkillable in the physical sense. The reason is just the topic of this book — the movement of life into space. As R.N. Bracewell has written:

"When we have colonized interplanetary space — which could be early in the 21st century, according to Princeton physicist Gerard K. O'Neill's timetable — we will have concomitantly achieved independence of the terrestrial catastrophes that lie ahead. Survival of the fittest, on a time scale of geological upheaval, may mean that communities over a certain age will be those that have succeeded in colonizing space."[19]

I would add a remark to Professor Bracewell's comment. Freeman Dyson has pointed out that there may well be very intelligent civilizations which have no interest in technology. I quite agree, but would guess that any civilization which becomes interested enough in the natural sciences to develop radio astronomy will achieve, almost at the same moment in its evolutionary history, liberation from its parent planet. Logically, then, I do not believe that war or natural catastrophe will constitute, in many cases, the limits to the duration of a civilization capable of communication.

I do have serious reservations about the probability that a civilization capable of communication, and stable enough to have a long lifetime on the galactic scale, will in fact choose to communicate. I readily concede that my reasons could be excessively anthropomorphic. They are closely connected with my misgivings about the entire concept of Project Cyclops.

On our planet we have seen, again and again, the effect of the contact between a primitive culture and a more advanced one. Almost invariably the more primitive is shattered. The destruction may not be intentional; often it may not even be physical. Yet it occurs, because the values and the knowledge gained over the centuries by the primitive civilization become, overnight, of little value in comparison with what is available from the more advanced.

When I have considered the effect of our discovering, one day, signals from a more advanced civilization (note that it would be, almost certainly, millennia more advanced than we are because of our own position at the threshold of communication) it has seemed to me overwhelmingly probable that the first effect of the discovery, as soon as the excitement and the novelty have worn off a little, would be to kill our science and our art. What purpose to study the natural sciences? We already know that they are universal, so if a civilization now radioing to us is as many thousands of years ahead of us in knowledge as we are from the Neanderthal, why continue to study and search for scientific truth on our own? Gone then is the possibility of new discovery, or surprise, and above all of pride and accomplishment. It seems to me horribly likely that as scientists we would become simply television addicts, contributing nothing of our own pain and work and effort to new discovery.

In the arts, music and literature, the case may be somewhat more unclear. Yet on Earth the almost invariable consequence of contact between a primitive civilization and one more advanced is the stagnation of the arts in the former. Only in the form of a "tourist trade" does art survive, in most cases.

If this sequence of effects is of more than local significance, as I think it is, it will be quite obvious to any civilization more advanced than our own. I would then add one more assumption: that the same characteristics which render a civilization immune to intellectual decay and stagnation, if there be such characteristics, are accompanied by a repugnance to inflict harm on others, in particular to other "emerging" civilizations more primitive than its own. In that case, "They may be out there, but they're kind enough to keep quiet."

If civilizations combining great age, great social stability and a continued lively intellectual interest do exist, and if those characteristics are accompanied by a concern for the development of primitives such as ourselves, is there any kind of signal that could be sent out to us that would carry great potential for good, and little for harm? Perhaps there is: the flash of a lighthouse, a simple message endlessly repeated, carrying just enough information so that we know it was formed by intelligence. The simple fact of its existence, proclaiming "you are not alone," could be of great help to us in our darkest moments. It would pull us outward, and spur our development — after all, we don't wish to appear as country bumpkins when true contact is finally made. At the same time, after the ten thousandth repetition of the same brief message, it will be clear to us that we must continue to gain our knowledge of the universe step by painful step, through our own efforts, and that physical travel to great distances will be needed before we answer the question: are they still there, or do we hear only the echo of a civilization that vanished long ago?

Proceeding now to the most speculative assumption of all, I consider that this age of natural science, in which we now find our own human civilization, may be a relatively brief epoch in the history of a long lived species. We are in the midst of a knowledge explosion, and if our rate of acquisition of new scientific knowledge continues to accelerate, as it is now doing, it seems to me quite likely that within much less than a thousand years we will know, if not everything about the natural world, at least so much that science will no longer be of great interest and challenge. In that case I would expect that our most talented individuals, a few of whom now study the natural and biological sciences, would turn their attention to the arts, or to the greatest intellectual problem that is now imaginable to me — the riddle of consciousness. My picture of an advanced civilization is one in which science, aided by computers with an intelligence level far higher than that of any living being, will already have answered all the merely physical questions. Some individuals may take part in direct exploration and exploitation of new star systems, slowly spreading the culture of their species in an expanding sphere from their parent star. I consider it probable, though, that in the advanced stages of a long lived civilization, the physical world will be taken for granted as something long since understood and thoroughly tamed. Most of the interest and activity, I would guess, will be intellectual, artistic and social.

After so much that is speculative, it's almost a wrench to return ourselves to the "little" world of our own solar system and the few decades immediately before us. We do so to consider the practical question which remains when debate about the value of Cyclops has gone on long enough: if it is going to be done at all, what is the best and most economical way to do it?

The answer to that question seems rather clear. Cyclops, in its original form, was studied by a group of two dozen people during the summer of 1971, at the NASA-Ames Laboratory in cooperation with Stanford University. The leader of this group was Dr. Bernard Oliver, of the Hewlett-Packard Corporation, and the result of the study is a thorough, excellently prepared report entitled *Project Cyclops*.[20] The report concluded with a proposal to construct, somewhere in a lightly inhabited desert region, an array of up to a thousand radio telescope antennas, each of large size, all of which would be steerable so as to point in a fixed direction as Earth turned under them, all braced against wind and storm, and all tied together electronically to function as a single giant receiver.[21] The total cost of the effort, if in fact all the antennas were built before an intelligent signal were received by them, was originally estimated as fifteen billion dollars. It could be reduced if advances in receiver sensitivity allow the same result to be achieved by a smaller array of antennas.

As an exercise, I looked into the question of building the equivalent of a Cyclops array as one of the early tasks of Island One. The space borne Cyclops would be far simpler — probably a single giant parabolic dish antenna, five kilometers across, located at a short distance from the space community. It would require only a single receiver system, which could easily be updated to remain at the summit of the electronic art as the years of search went on. The problem of noise arising from the many communications transmitters on Earth and in space would be overcome by the simple expedient of locating a disc shaped baffle, twice the size of the antenna, a short distance away.

For operation in the zero gravity, wind free environment of space the antenna and its noise shield could be light in structure, composed (in my estimate) of a geodesic frame covered by a thin skin of aluminum. The total mass including the shield would be hardly a tenth as much as that of an SSPS. Assuming that all of the complicated machinery (electronics, motors, motor drives, and so on) were brought up at high cost from Earth, and putting in generous figures for the costs of fabrication and assembly, the total expense for the Island One Cyclops still comes out only about one tenth to one twentieth of the cost of an equivalent installation on Earth. The L5-Cyclops would have a further advantage illustrated by an amusing little speculation: suppose that among the one million stars that are searched during a thirty year period there is in fact one which is beaming signals toward us; suppose further that the "program" has a duration of many years. After all, the beings at the sending end might be far more long lived than ourselves, and they might have a lot to say. Once we lock in on the signal, the L5-Cyclops can continue to point at the right place for as long as the program lasts. In contrast, a Cyclops antenna array on Earth, on the Moon, or in low orbit would be blocked half the time from receiving the signals. If that were the case, we can imagine the resulting Congressional investigation:

Senator X:  "Do I understand, Professor, that we are missing half the program the Arcturians are sending, and that you are proposing that we now build a new antenna system at L5 in order to replace the one in Nevada?"

Professor Z:  "Yes sir, that is quite correct. Of course, when the Cyclops program was initiated, we did not anticipate receiving signals with this particular time structure."

Senator X:  "Are you trying to tell me that when you came in here to ask for fifteen billion dollars, you weren't anticipating the possibility that your search might succeed?"

We will leave Professor Z in his rather sticky situation, and turn now to another application of the industrial facility at L5.

If the calculations I have described are not wildly off, the work force at Island One will be in a location so favored for industry that there will be strong pressure to enlarge the "beachhead in space" by constructing larger habitats.

Whatever group builds the first community, the success of Island One will prompt others to share in the earnings of L5 industry. Even on a three shift basis and with a population of which most people will probably be among the work force, the first Island One will not be able to satisfy, by itself, the demands which an energy hungry world will make of it. Even while the first few communities are being built, their designers (or possibly an entirely different group) will be planning for the next step in size: Island Two. The choice of size for the new generation of space communities should be made carefully, because for lowest cost it will be best to choose an optimum size and then replicate it in large numbers, using automated machinery and dry docks all suited to one set of dimensions.

Island Two should be large enough to form an efficient industrial base, but small enough so that transportation within its valleys will be easy, and so that its government can be simple and non-bureaucratic, functioning with a minimum of red tape. At a rough guess, the space residents may be ready to tackle something the size of Island Two after there are already a dozen or so communities of the size of Island One.

For economy of the structure that must contain the atmosphere, the internal pressure may be chosen similar to that in a high altitude town on Earth: Denver or Mexico City. Much calculation will be required before we know the optimum size for Island Two, but my present guess would place it not far from an 1,800 meter diameter (about 6,000 feet) with an equatorial circumference of nearly four miles. Island Two could house and maintain a population of 140,000 people, possibly in a number of small villages separated by park or forest areas. Each such village could be similar in size and population density to a small Italian hill town. As a former resident of such a community, I can confirm with nostalgia that it is one of the most pleasant arrangements for living so far developed on Earth.

Any comment I make about the city architecture and geography of a space community is, of course, no more than a guess. It may well be that several different types of arrangements will be worked out, perhaps even within a single habitat, so that without leaving the habitat people can enjoy the variety of "evenings out" in villages quite different from their own.

As in the surroundings of the first habitat, all the heavy industry of Island Two will be located outside, at least a few hundred meters away, in zero gravity.

Even while the first islands in space are being built, work will go on to upgrade the lunar mass driver for the task of increasing production of export products and of additional communities. A solar power station might be located on a mountain peak at the lunar north or south pole, where sunshine would be available full time. A transmission line from the pole to the lunar mine would allow the mass driver to double its throughput, without any change in the machine itself.

By the time construction of Island Two begins, there may be more than one mine on the lunar surface. Perhaps by then there will be a small industry on the Moon for building mass drivers and their solar power supplies. In the long run that will be the way to reduce shipping costs to a very low value (a few cents per kilogram). By then, too, we may be exploiting the vast reserves of the asteroids, and not long afterward, if the economics are favorable, we may shut down the lunar mines, and leave the facilities there as ghost towns.

The quite conservative economic scenarios that were developed early in our study of space manufacturing were based on a doubling time of about four years for the number of Island One communities, so that after about fifteen years the population in space would be over a hundred thousand. That number should be in the right general range to satisfy all the U.S. demands for new generator capacity soon after the turn of the century. It seems likely, though, that by then, if not earlier, the space communities will be responsible for supplying new generator capacity to all nations in need of it. As a rule of thumb, an Island One community, fully employed in heavy industry, could produce about 200,000 tons of finished products each year — more

Mining the asteroids should be even more advantageous and profitable than mining the Moon. They include the carbon, hydrogen and nitrogen missing from the Moon.

than two power stations, if it had no other employment. World needs by the beginning of the next century may be as high as fifty or more big SSPS stations coming on line every year, so the time may be not many decades off when the population in space may exceed a million people.

If automation is carried far, so that repetitive operations are done by a small work force, the replication time for habitats even of Island Two size could be as short as two years. The conditions at L5 seem made to order for such a development: zero gravity for assembly of large objects by lightweight machines; no weather, so computerized production will not have had to cope with the vagaries of seasonal variation and natural hazards; unlimited energy; and a task which consists of the repetition, thousands of times, of the same assembly operations with identical simple structures.

If the fastest possible time scale is reached, fifteen years from the beginning of construction there can be many communities at L5, with several hundred thousand people living and working in space. I hope that they will include young and old, children and the elderly, as well as people of working age. During those years, the sale, lease or donation of Island Two structures as "turnkey" industrial facilities appears to me a likely possibility. The cost of such a habitat should be not much more than that of the original Island One, because by then the work force at L5 will be quite large enough to produce every necessary variety of machine and part for construction. Only liquid hydrogen and possibly nitrogen or carbon will still have to be brought up from Earth.

For an underdeveloped nation or consortium of nations, the period during which several Island Two's are built will almost certainly be a time of excitement and opportunity. For a nation of a billion people (within two or three decades there will be at least two nations of that size) a space community for 140,000 people could then be bought, over a ten year period, for a cost equivalent to a few dollars per person per year. As a beachhead in space, from which further rapid expansion could then take place without additional foreign capital, such a community could be an attractive investment, especially when the possibility of its exponential growth is considered. It's speculation, and possibly nonsense, but thought provoking that with a replication time of two years for new habitats, a nation of a billion people, which purchased an Island Two Structure, would obtain in just eighteen years a growth rate of new lands in space fast enough to absorb a population growing even at 4 percent per year. Later I will explore that possibility further. For the moment, though, let's turn to what life might be like for the pioneers who may settle Lagrangia.

# Chapter 10
# *Trying It Out*

During the settlement of our own New World here in the western hemisphere, communications across the sea between family members were very important. Letters from the early immigrants allayed the fears of relatives left behind, and in many cases encouraged them to follow. In the settlement of L5 the communications with the "Old Countries" will be far more rapid — television phones can operate with a time lag of less than two seconds. It seems likely that even the earliest space communities will be equipped with electronic mail transmission systems, and I suspect that letters between family members will be as important for the humanization of space as they were to the settlement of our country. When time doesn't press, and there is a need for the comfort of handling the actual piece of paper touched by the sender, mail sent on a space available basis may be more satisfying.

Here are letters of a kind that might be written by people emigrating to L5 a few years after the first pioneers. Unlike those youthful emigrants who might be in the majority, this is imagined to be a couple whose children have grown up, married, and established families on Earth. Work experience and a record of stability and responsibility could be important factors influencing the "selection committees" which will play an important role inevitably in determining who goes to the early communities, though with the passage of time it can be expected that eventually most of the people who may wish to go to L5 will have the opportunity of doing so.

*Dear Peggy and Arthur:*

*Jan. 15, 20–: Jennie and I have been in Station One for twenty four hours now — I'll send off this note by the video mail while our impressions are still fresh. We were glad to get away from the slush and the wind up North, but even Cape Canaveral was pretty cold, and we heard that they were getting worried in Florida about the orange crop. Once in the Space Terminal it seemed like familiar ground to us, because of our six months at the Training School. Some of the people from our class were going to be on the same shuttle flight, too. After all the ticketing, the last medical checks, and getting our personal kits weighed, we went through to the locker rooms where we had to say good-bye to our clothes. Then showers and hair washing, and out the other side to the "clean rooms"; nobody wants to give a free pass for L5 to any plant eating bugs. Our space clothes were waiting for us, all clean and pressed. Of course we'd tried them on at the school, and Jennie'd sent hers back a couple of times to get the fit just right. I don't blame her — these light clothes don't leave a lot to the imagination.*

*The shuttle was already on the pad when we came into the waiting room; the spaceport crew was fueling up. We waited about an hour, but didn't call you — nothing new to say yet. Then the 150 of us filed on and settled into the bunks — the cushions were thin, but we knew we'd only be on them for half an hour. On the TV panels we could see our own lift-off, and it sure felt different knowing we were on top of those fireworks! The G forces weren't bad, especially lying down; just like the centrifuge at the school. At the end it was about three gravities, and I could still lift my leg without much trouble. Zero-G was very strange at first, but we kept still like the book says, and didn't get sick. On the TV we could see the shuttle moving up to Station One, and we felt the bump when we docked. The station hostesses floated in and helped us out and into the station — that took a while, about another twenty minutes I guess. Altogether, from lift-off to Station One was less than an hour.*

*They have a ramp leading "down" to the outer rim, so as we walked down gravity built back up to normal. The Station One lobby and restaurants have been on TV so often I won't say much about them, but I do want to tell you about the people. We were pretty lucky — there were only another 24 hours to go in the three day cycle between ships, so the station was pretty full. The seven shuttle flights before ours had brought up groups from a lot of different places: Chinese, Russians, a fair number of Indians, and one Nigerian group. From where I'm writing I can see Jennie — she's in one of the garden rooms and seems to have struck up a conversation with a girl who looks to me as if she came from someplace in South Asia — I guess they're both flower crazy.*

*Jan. 17: By the time all 2,000 of us were in the station the hotel was pretty crowded. It's good though that they have so many observation windows — we spent most of our time just goggling at Earth. I took a lot of color slides, because it may be two or three years before we get this view again. The hotel rooms were nice enough, but we didn't spend much time in them — too much to see, with Earth and the continuous movies they run in several theaters, and all the strange people.*

*We went to our room to see the Konstantin Tsiolkowsky come in — the view was a lot better from the TV. It was a pretty sight. First the end of the engine came in view, with its bright searchlight lighting up the clouds of vapor as they came shooting out. We could almost count them — like a movie running a bit slow. It took a long time for the whole length of the engine to go by, then the long straight mast, and the yardarms with their flashing red running lights. We couldn't see the guy wires that keep the whole thing straight. Then the ship herself came in sight: just a big ball, with no windows at all — like the head of a tadpole, and beyond it a big dish reflector for solar power. It took about three hours for all of us to file on with our kits and get settled into our cabins. We're not used to zero-G yet.*

*The captain gave us a nice speech over the video while the Tsiolkowsky got under way. Told us about how the passengers and crew are all on three shifts, matching three time zones on Earth eight hours apart: Moscow, Cape Canaveral, and Western Pacific. So the restaurants are going pretty much full time. There aren't any windows, of course, because of the cosmic ray shield, but the big video panels in every room give us good views, and they've got it set up so the cameras are fixed — you couldn't tell by looking at the video that the Tsiolkowsky is rotating.*

*Jan. 18: They sure keep you occupied here. I can see why the captain calls the Tsiolkowsky and the Goddard the "Flying Schoolhouses." Jennie and I are in classes for brushing up on our 800 word Basic Russian and Basic Japanese, and they've got a kind of nice arrangement about the meals. While one shift is having breakfast the other is having supper, and at that meal they use place cards. It's pretty clear what they're trying to do: sitting at a table for four, with a couple from maybe Russia or Japan or China, you can hardly help getting to know people. The Japanese couple we met this morning are in power plant construction, like us: he's an expert on casting titanium turbine blades. Since Jennie's been training for the last half year to be a blade inspector, they had some shoptalk to get into. The Japanese girl is an agricultural specialist, so I learned quite a bit about how they get so darned much food out of a little land up in the Japanese communities. I'll have to admit, though, that their English is a whole lot better than my "Basic Japanese." I think they're cheating — they use a lot more than 800 words!*

*There was big excitement early today when we passed the Robert H. Goddard on her way in toward Station One. She was in view for more than an hour, and our crew gave us some nice telescopic views, with enough warning so that there was a lot of camera clicking. The camera population must be more than the passenger list.*

*We met an Indian couple at dinner tonight. He's in construction, which makes sense I guess because the Indian government is concentrating on a fast buildup in the number of their habitats, rather than building mainly power plants like everybody else. We passed the orbit of the power stations on the first day — guess I forgot to say. Every now and then, as we spiral out, we get a bright glint off one of them, far in close toward Earth.*

*The L5 communities are getting visibly close now, and everybody is pretty excited. I've got to admit I get some butterflies sometimes: this is a young crowd, mostly, and even after all those tests I wonder sometimes whether Jennie and I, at fifty, are still good at learning new ways. So far we've liked it, I'll*

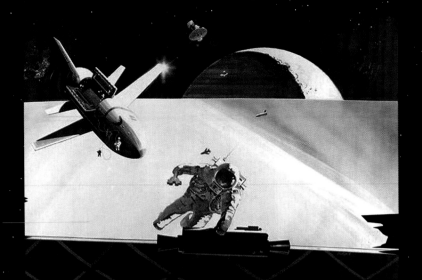

Since the late 1960's and the success of the Apollo program, the technology has existed to place men and materials on the surface of the Moon and in orbit around both the Earth and the Moon. Construction in space has been within our capabilities for more than a quarter of a century. Mankind has the ability to realize all of the riches that space can bring us — all that's been missing is the commitment.

Designs for space habitats have been in existence since the mid-70's and have undergone rigorous professional evaluation to verify their validity. The *Island One* design shown below could comfortably house 10,000 people and all of their needs. The colony in turn supplies energy and materials from space to better the lives of many thousands more people on Earth. And this could all be done with *Apollo*-era technology and for less than the cost of the *Apollo* program.

The *Island Three* design is a large cylinder providing three valley-like habitation areas, each 2 miles wide and 20 miles long, and each opposite a large window-like "solar" which admits sunlight. The sunlight is reflected on to the solars by variable mirrors to provide a "normal" day / night cycle. Each *Island Three* is circled by a collection of small agricultural habitats that provide all of the food for the 10 million inhabitants. *Island Three* habitats are well within the technology of the year 2000.

The living and working areas of a space habitat will not be the sterile metal boxes you might envisage. Landscaped, uncrowded and possessed of Earth-normal gravity, they will provide a living and working environment far better than what most of our world enjoys today. Industrial activities performed by an *Island Three*'s inhabitants will take place outside, preserving the habitat's unpolluted environment.

Agricultural areas, whether in the banded torus area of an *Island One* or in the external cylinders of an *Island Three*, will be mostly automated and will practice proven technologies such as multiple cropping and double planting. Because the environment and sunlight cycle of each agricultural area can be independently controlled, each area can be in its own "season", thereby making available fresh fruit and vegetables all year round. Agriculture in space will involve only those chemicals and insects deliberately chosen to be present so that crop yields are maximized and losses are minimized.

Gravity in a space habitat is provided by rotation. At the outer edges is Earth-normal gravity, but along the central axis (and outside) is zero gravity, where industries can be undertaken, and sports and entertainments can be enjoyed, that would not be possible on Earth.

Many aspects of habitat construction will employ a "modular" approach. Using the same prefabricated components repeatedly allows for both economies of scale and maximum automation in both manufacturing and assembly operations. This will place no constraints on the interior of the habitats — each area's functionality and appearance will be determined entirely by the needs and preferences of the people occupying it.

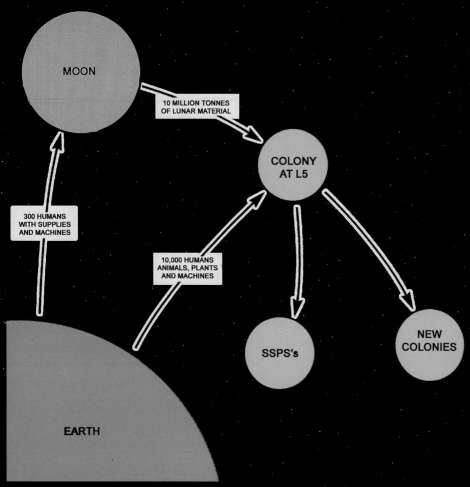

MOON

10 MILLION TONNES
OF LUNAR MATERIAL

COLONY
AT L5

300 HUMANS
WITH SUPPLIES
AND MACHINES

10,000 HUMANS
ANIMALS, PLANTS
AND MACHINES

SSPS's

NEW
COLONIES

EARTH

Space habitats will be built at "L5". Originally designating a specific point in space (the same orbit around the Earth as the Moon, but offset by 60°), L5 has come to mean any stable orbit around the Earth that's far enough from both Earth and the Moon that neither will eclipse sunlight from the habitat. This allows solar energy to reach the habitat full time.

The primary industry of the L5 colonies will be the construction and operation of satellite solar power stations (SSPS's). Solar energy available full time in space will be converted to microwave frequencies for transmission to Earth where it will be distributed to satisfy the ever increasing demand for energy — and thereby increase the standard of living throughout the world.

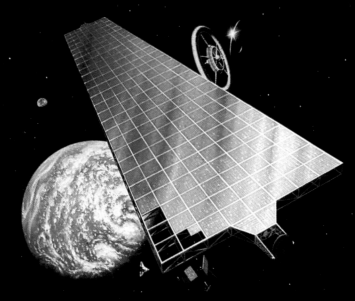

The Moon is a major source of raw materials, which can literally be scooped off the surface. An electromagnetic mass driver can quickly and inexpensively put these raw materials into lunar orbit where they can be collected for processing at L5 using free solar energy.

## LUNAR SOIL COMPOSITION

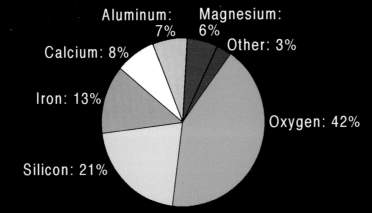

Aluminum: 7%
Magnesium: 6%
Other: 3%
Calcium: 8%
Iron: 13%
Oxygen: 42%
Silicon: 21%

The raw materials available on the Moon almost read like a shopping list for the materials needed for space habitat construction. Additionally, there is oxygen for life support and rocket propellant, and silicon for glass and for Earth's semiconductor industry. Only carbon, hydrogen and nitrogen are notably absent.

An even better source of raw materials is the asteroids — many of which cross the orbit of the Earth and therefore can be easily reached. The gravity of an asteroid is so weak that neither a rocket nor a mass driver is needed to lift materials from its surface. Asteroids will also be a source for carbon, hydrogen and oxygen, rather than lifting them from Earth.

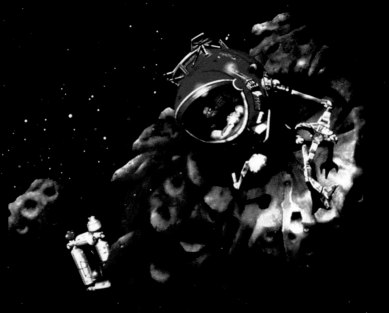

The current Earth orbital vehicle is the Shuttle. Although effective for putting small payloads in low orbit, it must carry extra fuel (at twice the payload's mass) to reach high orbit. The external fuel tank from each Shuttle launch is currently being thrown away to burn up. These tanks could easily be collected to provide raw materials in space, and could even be used in the shell of an early space manufacturing or living habitat.

The Shuttle is not the answer for high orbit operations (which is where almost all space activity, present and future, is to be found). Development is on-going to design vehicles that are fully reusable (unlike Shuttle launches, nothing will be expended) like the X-34 space plane. Also on the drawing boards are designs for "heavy lift" vehicles (HLV) that can lift large payloads into high orbits. The Shuttle's main engines and external liquid fuel tank are a definite step in the right direction, but more work is needed.

Human colonies in space — not a luxury, but a necessity. Earth is overcrowded, running out of raw materials, in desperate need of a growing energy supply, and being ecologically destroyed. The problems are worse with each passing day, and there are no solutions to be found on Earth itself. Mankind's destiny — its very survival — is in space. The materials, the energy, the room to live and our only hope are to be found simply by looking up. But a commitment is needed, a decision to go for it and the determination to see it through. The high frontier isn't just for a few highly trained specialists — it's for all of us. But we have to make it happen — you, me, all of us. Nobody's going to do it for us. Our parents handed us a pretty good world, but it's a world from which we've now taken all that it can give. What kind of world are we going to give our children?

*say that. The captain, in his little daily speech, is pretty funny. I guess he's developed quite a line of patter, after making the trip once every twelve days for a couple of years. I don't see, though, why he keeps apologizing about the food. It's a good deal better than what the airlines give you, at least. Today Jennie ordered a curry from the Indian menu. I copped out — took a ham steak, but I tried a bit of the curry and liked it.*

*Jan. 20: It was really great having that long video call with you this morning. That free half hour every week is going to mean a lot to us. Seemed as though the grandchildren had grown even since we left. Of course, we forgot most of what we wanted to say, and so much has happened that we couldn't have squeezed it in anyway.*

*We told you they docked us at Island One. They seem to use it partly as a receiving hotel now, a place where the Tsiolkowsky and the Goddard can dock, and where people get sorted out and sent to the communities they'll be living in. We exchanged some names and addresses, and already have some invitations to go visiting when we all get settled in a bit.*

*Island One is small, of course — only 500 yards in diameter. This one runs at Canaveral time, and two other communities of the same size, nearby, run on the other time zones. Many of the people we met on the ship landed at the others, so as not to be time shifted on arrival.*

*I wonder what it was like for the people who lived in this first habitat for several years, before the first of the Island Two's were done. Not too bad, perhaps. Jennie and I were given one of the smaller apartments, two big rooms with kitchen and bath, with a nice garden. This first Island One runs at a constant Hawaiian climate, because they weren't sure about temperature changes in the beginning, and didn't want to take chances with the structure. The old timers say the climate in One is dull, but after Michigan in January we're happy to soak up the sunshine for a while. The garden has some big tropical flowers, and I can see that the people who lived here first liked avocados; they had several trees, and one was just right to give us a fresh avocado with our lunch.*

*It really is sort of a vacation atmosphere — the good weather, and so many new things to see. Of course, we had to try Island One's first new possibilities: human powered flying and the slow motion diving.*

*On our deck chairs in the garden, soon after we settled in (I won't say "unpacked," because with a 100 pound baggage limit, we didn't have that much to unpack!) we could look up and across at the big corridor that leads to the "ag area" — the part where they grow the crops, using all mechanical equipment. The curved surface of the habitat is all terraced and planted — a mass of green and bright colors. The Sun is at an angle of about 11:00 AM, all the time except when they shut it out with some screens outside, for nighttime. Each morning around seven there's some "rain," so when we wake up everything is fresh and there's a nice clean smell of rain and flowers in the air. Island One is too small to have any real weather of its own, though, so the "rain" just comes from some spray pipes that we can see, seven hundred feet above us, when we look hard.*
*Exactly above us we can see the gardens of the apartments on the far side, and then the curve of the sphere. For some reason, it doesn't seem so strange to us to have trees growing straight down as it does to have them seeming to come out horizontal, as they do from the gardens a quarter circle away from us.*

*Many of the gardens are open, but someone told us that the settlers who prefer the small size of Island One, and have stayed here rather than move, have little gauze sunshades over part of their grass, so that they're able to sunbathe in the nude and still not be seen from the "sky." As we look up we can see the figures of people flying, 800 feet above us.*

*The apartments here are grouped in terraced buildings, so that each can have its garden, and the buildings are set into little villages, separated by trees and parks. In our exploring we're getting our exercise, because there are pathways, not roads, and most of the villages are at least part way up the slope from the "equator."*

People seem to love flowers here; all the paths are lined with them. I guess it's just so easy to grow them — no weeding, no bug spraying, and just the right amounts of rain and sun. I understand that in each community the Garden Club is one of the most important organizations there is, and that particular people volunteer to take care of individual little areas of the paths and parks.

Down near the river there's "Fifth Avenue" where nearly all the shops are. It's on two levels, with wide sidewalks and a lot of planting. About half of it must be little restaurants — it seems that when Island One was new everyone was working so hard, men and women both, that a lot of them didn't do much cooking. The restaurants are all small, and many of them are buffet style, most with salad bars. There are numerous book shops, and a main library, and quite a few small cinemas.

Still lower down, past a belt of trees, there are tennis courts and the playing fields, and of course the park and beaches at the equator itself, by the river.

On our first exploration we kept looking up and seeing people flying, and it looked too good to miss, so we began walking up through one of the villages, and beyond it to where the hill gets steeper and steeper. The sensation was really strange, because as we climbed we got lighter. Past the green park area we were on one of the bridges across the windows, climbing at more than a 45 degree angle, but finding it easier because of weighing less. Beyond the windows we were on a very steep, winding path, through almost a jungle of ivy and shrubs, like a Hawaiian hillside. Up at the very top we found a number of people, because all of us newcomers were exploring at the same time. I'll have to admit that as soon as I tried flailing around in the zero-G clubroom (it doesn't rotate) I started feeling a bit ill. Jennie loved it, though, and when someone had a pedal plane free she was the first of us to try it. I watched her from the zero-G room; the pedal plane sets you at an angle almost like lying down, and there's only a bar at waist level — no real seat. The wings are small, but there are three sets of them; it's a triplane. The two big propellers are almost as big as the wings, and go in opposite directions when you pedal.

Jennie had some problems right at the axis, because there wasn't any "down" and the plane was designed to work in at least a little bit of gravity. Once she moved down a few feet, however, she was OK and pedaled out. She stopped pedaling and drifted into the thin netting that's out there, pedaled for another quarter mile, and then turned around to come back. Then came the problem — she was getting tired and it seemed quite a distance to go. She settled down and rested on the netting for a few minutes — out there it's attached to the "rain" pipes. By the time she got back I was feeling better and took a short flight myself. I didn't try to go all the way to the "South Pole," though. Somebody told us that in Island Three there's going to be a cocktail bar hanging in space at the .05 G level, half a mile out from the end. That flight will get people's thirst up!

Feb. 2: Your Dad is starting to get deep into his work, as he always does, so the vacation is over for a while and I'd better take up the letter writing. It was really thoughtful of you to make such nice preparations for our weekly call. The children looked great. I think they're getting used to the calls and aren't so shy any more.

At the docks on our way over to Island Two we met a rather sad group — some of the people from our training class, who've decided that they just can't take it here, and are going back to Earth. It's not a physical problem, because almost no one feels any dizziness in a habitat as big as Island Two. I guess that for those people all the newness and differentness are just too much, and they haven't been able to get over it. The old timers tell us it's a well known thing, and they call it the "wide syndrome." We didn't know what they meant until someone tipped us off: it's WAIDH, for "What am I doing here?"

After Island One, "Two" seemed really big. The basic layout and the landscaping are similar, though, except that "Two" is not so warm, and runs with a climate that's right for pine trees and firs. You know

*I like rhododendrons, and our apartment has a lovely mass of them against the garden wall. I don't think that Dad told you much about our apartment, but you could see some of it from the video. They've done a good thing here: because the Sun is overhead, almost, they've been able to fix it up so that there's a space of a foot or so between the apartments, and the sunshine comes down that and shines on a planter that's just outside a long window of the living room. It makes the living room sunny, and gives us very good soundproofing — we can't hear the neighbors at all. The birds are fairly noisy, though, especially in the morning just after the rain, when the butterflies first come out.*

*We sample the restaurants quite often, and meet people that way. They're a nice sort, and we feel very safe and secure here. Maybe because we all arrived with so little, and because most of our salaries are being put in the bank, no one seems to lock his doors. I like getting my groceries at the supermarket; that you'd really find different! The vegetables and fruit are spectacular, especially the tropical ones. At first I felt like buying up strawberries and guavas, but we're getting used to the fact that they're "in season" all the time here. Dad misses his steaks, but I tell him that back on Earth we could hardly ever afford them anyway, so he shouldn't complain. I've joined a cooking club, and the Garden Club, and am trying to duplicate a recipe we had in one of the restaurants last week. It's chicken, but cooked almost with a taste of seafood, like lobster. Next I'm going to cook a glazed ham, because with two people we can get a lot out of that for a week or more.*

*We both like the low gravity swimming, especially the diving. The water comes up to meet you so slowly that you have plenty of time to do two or three flips before you meet it.*

*One thing we both really like is the six day week with only four working days. I say only four, but really there are so many clubs and volunteer jobs that we find ourselves working harder on the weekends than at our plant. Of course, it's set up this way so that the parks, the restaurants, the churches, and all the rest of the facilities get used efficiently and without crowding. With only one third of the population having a weekend at any one time, you don't find the parks empty one day and crowded the next.*

*Feb. 15: I could never get Dad to a ballet back home, but the Russian company from one of their communities was here last week, and we both had to see them. It was in one tenth gravity, of course, and we both realized that ballet was really meant to be done that way. I don't know all that much about it myself, but anyone could see that all the easiness, and lightness, and the whole dreamlike quality of ballet is just so much better without gravity pulling down every motion. We came away just stunned.*

*I'm addressing this one just to you, Dear, because much as I love my son-in-law there are some topics I'd feel shy about with him reading the letter. All I can say is, I hope that you and he get a chance to come up here one day. We'd heard a few remarks about the zero gravity hotel, of course, but nothing we'd heard could have prepared us for what we found. Dad had it all arranged for our anniversary, but kept it a secret from me. First he took me to a really wonderful little Italian restaurant in one of the villages high on a hill: all candlelight and soft music, a terrace with a view, and good food. Then, with only a brief stop at home to pick up our things, we were off to the Floating Island Hotel for our weekend. Most of the hotel, like the lobby and restaurants — and the showers — are at one tenth gravity, but those bedrooms! My dear, it's just indescribable. Of course, you could watch TV or listen to music if you want, but really, as Dad says, those rooms are designed for just one thing. I can't imagine you two ever not getting along well together, but if you ever have a problem, before it gets too serious bring him up here for a second honeymoon! You may never want to go back. Now that we've found what it's like, I can tell you it's going to be a lot harder for us to leave!*

*With much love -*
*Contentedly,*
*Jennie*

In "colonizing" space, the emphasis is on providing opportunities and
an appropriate environment for large numbers of common people,
not just for small numbers of specialists.

The days in which Edward and Jennie voyage to their New World are taken to be twelve to fifteen years after the completion of Island One. On the fastest possible time scale, there might be at that time a rise of the total population in space from 500,000 to one million within a two year period. That's about seven hundred people each day — not much compared with the traffic through one of our major airports, but more than the space shuttle could cope with unless the fleet and the launch facilities were to be expanded greatly. I assume, in keeping with studies already carried out by NASA and its contractors, that well before the end of this century there will be shuttle vehicles, propelled by chemical rockets of a somewhat more sophisticated type than those of today, capable of lifting off Earth and accelerating to orbital speed without staging (dropping components) at all. Such single-stage-to-orbit vehicles are said to be within 1980's technology, so I don't think it's being rash to assume they'll exist by the late '90's or the early years of the next century. They'd bring the cost of Earth-to-orbit transfer down considerably. There is an often quoted observation by Theodore Taylor that's quite relevant to the question of present day space transport systems:[1]

(Paraphrasing):
   "Present costs for putting freight into orbit are high for the same reasons that jet travel on Earth would be expensive if the corresponding rules were followed for the operation:

1. There shall be no more than one flight per month.

2. The airplane shall be thrown away after each flight.

3. The entire costs of the international airports at both ends of the flights shall be covered by the freight charges."

The way to obtain lower costs for lifting freight into orbit is evident from this quotation — develop vehicles which are fully reusable, and find a market large enough to justify frequent flights. There are, though, two "catches" in this reasoning. First, the studies which have been made so far indicate that with chemical rockets it would be extraordinarily difficult, if not impossible, to build a fully reusable vehicle capable of making a round trip from Earth to L5 without refueling. Second, the development costs for any vehicle that requires a big leap beyond the existing state of the art are very high. For the so-called "super shuttle," for example, a vehicle capable of taking enormous payloads to orbit and of making round trips without discarding any of its components, I have seen NASA estimated development costs of $40 to $60 billion. The vehicle which I imagine in the letters of Edward and Jennie is of a more modest sort, carrying a much smaller payload.

During the time period about which I am now speculating, I'm assuming that it will not yet be practical to obtain carbon, nitrogen and hydrogen from the asteroids. For conservatism, then, it seems safest to assume that it will be necessary to bring up from Earth about one ton of those elements for each emigrant. Such freight would not need to travel on the same very safe vehicle that would be needed for human transport.

The limiting factor with the Space Shuttle lies in how much of the launch hardware is expended on each mission. Reusability is the primary concern with newer launch vehicles being designed.

For transport of seven hundred people per day, by single stage rockets with payloads only two or three times that of the existing shuttle, there would only need to be about five flights each day. For a completely reusable vehicle that doesn't require assembly, only to be fueled up before each flight, such a liftoff rate doesn't seem high, even if by then we do not have additional launch sites beyond the existing two (those of the United States and of Russia). Freight requirements, though, might be higher in terms of tonnage; probably not in terms of flights. A flight every three hours or so by a shuttle derived HLV would be enough to bring up the required supplies to initiate agriculture and to establish a comfortable environment even during the period of rapid buildup of population at L5. By the time we need that sort of freight hauling, though, we'll probably use the same single-stage-to-orbit vehicle to do the job. A few flights each day will be enough, and that vehicle will probably burn much cleaner fuels than does the existing shuttle.

The problem of travel beyond low orbit is quite a different one. The advantages of full time solar energy and easy access to lunar materials can only be enjoyed at escape distance, but to go from low orbit to a great distance requires a far longer time and, if Earth is still the source of supplies, relies on a longer and thinner supply line. The problem is analogous to that of an extremely long range aircraft flight. If we require that the plane reach its destination, turn around, and return without refueling, we make the problem far more difficult than if we permit refueling at the destination for the return trip.

The problem of low orbit to L5 transfer is, for passengers, first that of time. Even with high thrust engines, able to make large changes in the velocity of the rocket within a period of only an hour or less, the travel time to escape distance is about three days. The simple type of accommodations that would be adequate for a flight of half an hour or even of several hours would be quite unbearable for a trip lasting for a number of days. "Steerage to the stars" is not the image that we would like to look forward to in connection with the humanization of space.

The X-34 is the latest government vehicle under development and it may turn into a useful system in the effort to promote space colonization.

Fortunately, there are compensating advantages of which use can be made to get around this problem. From low orbital distance out, there is no requirement that vehicle engines be capable of supplying a thrust greater than the vehicle weight. If we are willing to settle for a slow trip, engine thrust and acceleration can be quite low. If we make use of the fact that L5 will be a site at which reaction mass will be relatively cheap, it seems clear that instead of developing monster vehicles for liftoff from Earth, we would be better advised to solve the problem from both ends. L5 is the ideal site for construction of large spaceships, whose design could be free of any of the limitations forced by entry into planetary atmospheres. Mass driver engines for those ships can "fuel up" at L5 with reaction mass either in the form of industrial slag or of liquid oxygen.

The spaceships *Konstantin Tsiolkowsky* and *Robert H. Goddard* are assumed to have empty masses of about 3,000 tons, of which about two thirds would be their mass driver engines and their solar power plants. The mass driver engines would have exhaust velocities about twice as high as for the best chemical rocket — about the same as for the much earlier but similar machines studied intensively in the late 1970's for the early days of space manufacturing. Those engines, carrying solar cell arrays like the sails on a square-rigger, would stretch out for several kilometers, but that would be quite tolerable for vessels never intended to enter an atmosphere.

To find the performance of the *Goddard* we have to know how much the solar cell arrays will weigh. I'm assuming three and a half tons per megawatt. The NASA Johnson Space Center, in a detailed study, concluded that it could do that well even by the 1980's, for a satellite power station. For the *Goddard*, years later in time, that should be attainable — especially so when we remember that for a spaceship engine there is no need to hold the cost down to the low value that would be required for an economical central power station. For the *Tsiolkowsky*, the *Goddard*, and their sister vessels the corresponding travel times would be around three weeks for the inbound leg of the journey, and just over a week for the outbound — about the same time that it takes to cross the Atlantic on a medium sized vessel. The differences in trip time arise from the fact that the engine would have constant thrust and that on departure from L5 each ship would be heavy with reaction mass. That difference would be a happy one for the outbound travelers, who would enjoy a higher average speed than would the crew when spiraling down to low orbit from L5. Later on by perhaps two decades, when transport requirements may be much greater, the engineers may be able to make still lighter solar cell arrays. If they can produce something in the ton-per-megawatt range, the travel time can be reduced to little more than three days. Other approaches, including the possibility of laser or microwave beamed power, are not out of the question. I am not considering the possibility of nuclear power. The reason is straightforward: if the development of the communities is to go on without check for a long period, one must not design into it "absurdities" that would pose a limit as soon as total numbers or total required transport exceeded some modest value. It does not seem to me to make sense to design a deep space transport system around an energy source that would have to come from Earth.

We can get lower and upper limits to the ticket price for a trip to L5 in the late 1990's-early 2000's time period. The lower limit comes by assuming round trip times of a month, and ship costs per ton that are three times as high as those of present day commercial aircraft. The total comes out around $6,000. The cost of reaction mass would only be a small fraction of that total, because it would be so abundant at L5. A still lower ticket price could exist if the ships carry full loads of either passengers or cargo both on the inbound and outbound legs of the journey.

The upper limit is $30,000, and comes by assuming that each vessel must collect in revenue an amount equal to its own cost, within a time of eighteen months. Ticket costs on commercial jets within the United States have about that ratio to aircraft purchase price; they include, though, total fuel costs which are a higher fraction of the cost of the capital equipment. Either the $6,000 or the $30,000 figure would be a small fraction of the productivity of an industrial worker in a single year at the favored location of L5, and would probably equal only a few months' earnings.

Edward, Jennie, and their co-emigrants are assumed to arrive on the scene after the first settlers of Lagrangia. The first arrivals will have faced more difficult conditions and a more limited environment, but one much less harsh than our ancestors faced when the New World was opened to colonization. There will be no "hostile Indians" and there will be plenty of food.

As noted earlier, agriculture at the communities will be intensive and highly mechanized. By the scheduling of opening and closing thin shades several kilometers distant in the Sunward direction there will be long summer days in the agricultural areas, with never a cloud or storm, all year. Monotonous, but a field of grain doesn't demand variety. Temperatures in the agricultural areas will be kept always very warm, so that corn, sweet potatoes, sorghum, and other fast growing crops can grow from seed to as many as four harvests in a year.[2] Cuttings and some grains will be eaten by chickens and pigs and turkeys, so that the settlers can enjoy a varied diet including high protein meats.[3] There will be no need for insecticides or pesticides in the Islands, because agriculture will begin with carefully inspected seeds from Earth. The initial lunar soil will be sterile, and nothing will be introduced into it but water, chemical fertilizer and necessary strains of bacteria helpful to plant growth.

Though beef cattle may be too wasteful of space and too inefficient in converting plants to meat to earn space in the early communities, there will be good reason to include a small stock of dairy cows: children will still need their milk.

In the villages of the Islands there must be insects, perhaps butterflies, for the birds to eat, but there need not be mosquitoes — or cockroaches, or rats. The space clothes described by Edward and Jennie can be light, because their environment will be free of harsh extremes. They will find no "wide open spaces," but then neither do the inhabitants of northern areas on Earth, who must pass long winters mainly indoors.

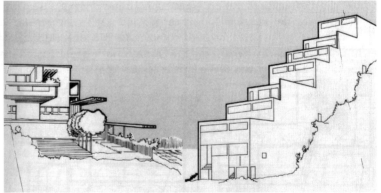

Because of the curvature in space habitat designs, terracing may be the best way to maximize residential land usage and provide direct sunlight for all dwellings.

In trying to guess what life will be like for the early settlers, we should recall one of the most deep rooted of our own needs as healthy human beings — the need to feel that we are of value, that our effort and work are needed and appreciated. In Island One everyone will have the sense that his work is needed and is important; there will be no unemployment. Probably the early residents will develop close-knit communities, and their villages may develop identities of their own even though they are no more than a few minutes' travel time apart.

Many of the enjoyments of the early communities will be those which we would expect in a small, wealthy resort community on Earth: good restaurants, cinemas, libraries, perhaps small discotheques. Yet some things will be very different. There will be no cars and no smog — travel will be on foot or on bicycles. By the time the first Island Two is built, it may be possible to have not just the small river imagined for Island One, but a lake. With inexhaustible solar energy close at hand that lake may have beaches lapped by waves perhaps even large enough for surfing.

# Chapter 11
# Homesteading The Asteroids

Our ability to send people far into space, and there to maintain them in good health, reached its limits with the Apollo Project. The life support systems developed for that venture were capable of maintaining human life for two weeks, long enough only for a quick dash to the Moon, a few days of exploration, and a return. The Skylab Project of the early 1970's extended the time limit for astronauts in space to three months, but in a location much closer to the Earth, in low orbit.[1]

The maintenance of life for a time of many months, as would be necessary for a voyage to the asteroids, presents no problems that are new in principle, but the detailed engineering of such systems has not been done. We should assume, then, that the early space communities will be built entirely of materials from Earth and the Moon. As the number and size of the islands in space increases, there will be a demand for additional materials to stock them. That demand will be a strong incentive to tap the resources of the asteroids for carbon, nitrogen and hydrogen, though it is likely that for some time all other elements will be obtained from the lunar surface.

A serious proposal for the humanization of space could not have been made before the Apollo lunar samples were returned to Earth. Similarly, it's only in the past several years that our information on the compositions of the asteroids has advanced from the stage of speculation to that of near certainty. The recent increase in knowledge and understanding has come the development of three new techniques[2,3]: the measurement with high resolution of the dependence on wavelength of the reflection from an asteroid of sunlight, in the visible and near-infrared region (spectrophoto-metry); measurement of the polarization of the sunlight reflected from an asteroid (polarimetry); and the measurement of infrared light from an asteroid (radiometry). The last two methods combine to give a measure of the diameter and the average coloring (light or dark) of each asteroid, and the first is now so sophisticated that spectra characteristic of particular minerals can be recognized in the light reflected from an individual asteroid.

Whether living in a colony at L5 or homesteading the asteroids, space habitats will have both Earth-normal gravity and zero-G environments. Just like on Earth, where we live and where we work are often quite different.

More than 90 percent of the asteroids fall into the classifications "carbonaceous chondritic" or "stony-iron." These classes correspond to groups of meteorites found on the surface of Earth. Carbonaceous material is not unlike oil shale, being rich in hydrogen, carbon and nitrogen. It is generally soft and friable, and can be melted at a low temperature. Probably for that reason, not many carbonaceous meteoroids survive their fiery passage through our atmosphere. In the more benign environment of the asteroid belt, much more of the carbonaceous material has survived. There is fairly good evidence that most asteroids are carbonaceous, including the two largest of those minor planets, Ceres and Pallas, with diameters respectively of almost a third and about a seventh that of the Moon.[4]

The energy interval between the asteroids and L5 is almost exactly the same as that from the Earth to L5. For a practical rocketeer, that energy interval is expressed by velocity changes that must be made in order to change the orbital radius and to tilt the plane of the orbit from that which matches an asteroid to that of Earth and Moon. Asteroids in the "main belt," out beyond Mars, move relatively slowly in their orbits. Earth, nearer to the Sun and therefore more strongly attracted to it, must travel faster in order not to be pulled deeper into the Sun's range of gravitational influence. The difference is typically six kilometers per second, and must be made up in the course of any voyage to or from an asteroid. Further velocity changes must be made to match the eccentricity (lack of circularity) of an asteroidal orbit. Most asteroids circulate in planes inclined to that of Earth (ours is called the "plane of the ecliptic"). For each two degrees in angle by which the planes differ, an additional velocity change of about one kilometer per second must be made. If we search through the list of asteroids for those with favorable orbits, and calculates the total velocity interval which separates each from L5, the answer in nearly all cases turns out to be near ten kilometers per second. The velocity interval between Earth's surface and L5 is only slightly higher.

Although these velocity intervals to L5 from Earth and from the asteroids are so nearly alike, there will be at least two incentives for obtaining carbon, nitrogen and hydrogen from the greater distance rather than the lesser. In deep space, high thrusts will not be required, nor will spacecraft hulls to protect payloads during their brief passage through Earth's atmosphere. In the long run, the economies made possible by those additional freedoms will almost surely tip the transport cost scale in favor of the asteroidal resources. That shift, when it occurs, will avert what would otherwise become an increasing burden on the biosphere of Earth from rocket flights through the atmosphere. Eventually, it seems likely that transport up to low orbit, and from there to L5, will be needed only for people and for particular products, especially those that are light in weight, but which can only be made by specialists from among Earth's large population.

The time delay before the asteroidal materials can be exploited may be estimated by the duration of the Apollo Project. In that case about eight years was required to progress from the early, primitive Earth orbital flights to the successful round trips to the Moon, a thousand times farther away. In order to go to the asteroids, it will be economically advisable first of all to have a well established facility at L5. The early space manufacturing communities can supply as by-products of the industries reaction mass for impulse engines. Those communities can also serve as shipyards for the vessels of deep space. By contrast, an asteroidal journey starting from Earth would require several times as much energy as from L5, and would involve the added expense of vehicles for precision lift off from the planetary surface under conditions of strong gravity. If the construction of an asteroid voyaging ship is begun several years after the first L5 community is established, the first human venture to the asteroids might begin within eight years from the dedication of Island One. It would be preceded by relatively inexpensive unmanned probes, also launched from L5, so that the first travelers would go to an asteroid that is already known to contain the elements wanted. The situation is quite unlike that of speculative oil well drilling into the surface of Earth. We can know more about the composition of an asteroid a hundred million miles away than we can know, without drilling, about Earth a thousand meters beneath our feet. In space there need be no wildcat oil operations or dry holes.

For economy, our transport system should be the analog of Earth's cheapest — a tugboat and a string of barges. In space, where there is no drag, we could practice a further economy. The tugboat would be needed only at the start and end of each trip, and during the long months of the orbit from the asteroidal source to the region of L5 the payload, in the form of tanks of ammonia and of hydrocarbons, could travel unmanned.

Like Earthbound tugboats, ours would be mainly engine and not very beautiful. One conservative design could be based on a longer version of the mass driver used on the Moon. It could be many kilometers long, if braced by yardarms and wires like the mast of a racing sailboat. The structure could be lighter if the payload were distributed at intervals along the whole length of the engine, because the thrust of the engine would be distributed equally along its length. The tugboat might be powered by lightweight photovoltaic solar cells, assisted by large, lightweight mirrors to concentrate the weak sunlight of a distant orbit. The active electrical components of the mass driver might be contained in a long, thin aluminum tube, pressurized with oxygen to the equivalent of a mountain altitude on Earth. That would permit maintenance and repair of all components likely to give trouble, without the inconvenience and lost efficiency that accompanies the use of space suits.

There would probably be living quarters for six or eight people, sufficient for three watches as on a boat at sea. There would be a small chemical processing plant, sufficient to form reaction mass from asteroidal debris. Altogether, the tugboat might have a mass of a few thousand tons, comparable to that of a large Coast Guard icebreaker on the oceans of Earth. The payload, in the form of tanks of chemicals, could be as much as the cargo of an oil tanker. After months of steady pushing the payload would have acquired the velocity changes necessary to put it on orbit to L5. The tug would then cast off, and return home to the asteroidal outpost community. The returning crew would take time off while the tugboat was piloted on its next trip by a rested crew. Meanwhile the linked payload would swing silently inward toward the sun on an eight month flight that would bring it to the vicinity of L5, and rendezvous with another tugboat for the final velocity change.

Both solar power and mass drivers will be employed near to Earth, out in the asteroid Belt, and to provide transportation between the two.

Tugboats on the oceans of Earth, even exposed as they are to storm and damage, often last fifty years. It has been one of the phenomena of the early years of experience in space that satellites usually last much longer than their "design" lifetimes. For the mass driver powered tugboats of the asteroidal belt, which would operate without high temperatures or pressures and would never be exposed to wind or storm, the lifetime would probably be much longer.

Probably they would be retired by obsolescence rather than by wear. Transport costs for material from the asteroid belt, based on present day figures for interest rate, amortization and costs of aerospace equipment, lie in the range of less than a dollar to several dollars per kilogram. That's far higher than the cost of supertanker transport on Earth, but much lower than the costs for any presently conceivable transport system operating from Earth's surface to L5.

As has happened so often when we've studied in depth possibilities that seemed promising as aids to space manufacturing, the asteroids may be even better sources of materials than I've suggested so far. Though most of the minor planets are in the main belt, Dr. Brian O'Leary pointed out that a special class, named after the asteroids Apollo and Amor, have orbits much closer to the Earth's. In a 1977 NASA-Ames study O'Leary gathered leading experts on asteroidal measurement and orbit theory. They worked out detailed scenarios for recovery of specific, known asteroids of the Apollo-Amor class, using mass driver reaction engines. Their technique made use of "gravity assists," and in action that would be spectacular indeed. After rendezvous with an asteroid, the tugboat crew would so direct their engine that the asteroid would swing by a planet like Venus or Earth. At the swing-by, the asteroid's velocity would be changed as much by the planet's gravity as it would by months or years of mass driver operation.

With the help of the gravity assist technique, already well proven in space probe missions to the outer planets, it seems that some of the asteroids may be much more accessible than those of the main belt, and from an economic viewpoint may even give the Moon a run for its money. There's plenty of material available — even the smallest asteroid that we can see in our telescopes has a mass of more than a million tons.

At a certain point in the growth of the L5 communities, trade between the islands of space will begin to dominate over the "colonial" economy of interchange with Earth. We have seen that transition take place in the colonies of the Americas, Africa and Australia. It seems likely that for any new community whose major purpose is the habitation and maintenance of its population, rather than of supply to L5 or to the Earth, economics will favor its construction without any prior shipment of materials at all: that is, in the asteroid belt itself.

The construction equipment for building a new habitat could be sent to the asteroids from L5, or manufactured in the asteroidal region. With that equipment new habitats could be built from material readily at hand, and as soon as each new habitat is ready its population could travel from Earth or L5 to occupy it. The saving in transport cost by that development would be significant. The weight of the settlers who would move into a new habitat would be only about one five thousandth of the weight of the habitat itself. Again, there's plenty of material. To build a colony the size of Island Two, for a population of more than a hundred thousand, an asteroidal chunk a few city blocks across would be sufficient — a mere speck, at the margin of visibility from Earth.

Once in operation, a space community would be quite capable of moving itself, in a leisurely fashion, to another point in the solar system. To do so in a manner economical of reaction mass might require a technology presently being studied, but not yet realized at the level of engineering practice — the acceleration of tiny pellets or grains of solid material by electrostatic forces. The ion rocket engine, a device already built and tested in the anticipation of scientific probes to the asteroids, works by the same principle; the difference lies only in the size of the pellet being accelerated. The ion engine would accelerate something more nearly like a grain of dust.

Until the intensive theoretical study of mass drivers in the late 1970's, they would not have been thought of as serious competitors to ion thrusters for high performance missions. Now, though, it appears that a mass driver might perform quite well even in the demanding assignment of moving a completed habitat through some great distance within the solar system.

During the development of chemical rocket engines, exhaust velocities have increased steadily. The higher the velocity of the exhaust, the less fuel need be carried for a given task. In the case of an ion or pellet engine, though, high velocity is not always desirable. The velocity of an ion, in the case of engine with parameters that permit easy operation, is so high that the performance is limited by the electric power available. If one halves the exhaust velocity of an ion engine, the reaction mass required to carry out the mission in the same length of time doubles, but the electric power required decreases to half of its previous value. For any given task there is an optimum exhaust velocity, just high enough so that the expenditure of reaction mass is not intolerable, but low enough to minimize the amount of electric power needed for the engine.

In the case of a moving island in space, the optimum exhaust velocity is five to ten times that of a chemical rocket, if the task is to move the newborn community from the asteroid belt to the vicinity of L5. For such a velocity the amount of reaction mass used in the trip would be only a quarter of the habitat mass. It would be obtained in the course of the voyage by processing a cargo of asteroidal rubble, possibly by a simple grinding and sieving operation. The lifetime of a community would be indefinitely long, given continuous habitation and maintenance. On a time scale of at least thousands of years, it would not seem unreasonable to devote thirty years to a relocation. Based on present day costs for turbogenerators,[5] the necessary power supply installation for that task would be capitalized at $25,000 to $60,000 per inhabitant, certainly not an exorbitant figure. In the last twelfth of this volume I will describe just how far a community could go, if possessed by wanderlust. For the moment, though, it is enough to point out that the choice of location might be made by a vote of the inhabitants, and that the choice might not always be that of returning toward L5. Any orbit within the entire volume of the solar system, out to a distance farther than that of Pluto, could be reached within less than seventy five years by a space community. Within that huge volume it would always be possible to obtain a full Earth normal amount of solar intensity, by the addition of lightweight concentrating mirrors to the light reflection system that an ordinary habitat would carry. A community or a group of communities desiring a peaceful and quiet life might well choose not to return toward Earth, but to "go the other way" to a private orbit from which the interaction with the population near the Earth would be, at most, by electronic communication.

We should realize that the humanization of space is quite contrary in spirit to any of the classical Utopian concepts. At the heart of each Utopian scheme, including the modern communes, there have nearly always been two very different, even conflicting ideas: escape from outside interference, and tight discipline within the community — freedom and constraint.

Escape from outside interference will be an option open to a community in space, unless military intervention occurs to prevent it. There will always be the possibility of "pulling up stakes" and moving the habitat to a new orbit far from the source of the interference. In history we have many examples of groups, not least among them our Pilgrim ancestors, who have been permitted to escape from coercive situations. Usually those who remain behind justify that permission by something equivalent to "We're better off without those troublemakers." The space communities would be in contrast to the classical Utopias in part because they could escape so much more successfully. Here on Earth the possibilities for escape are limited because a community that desires isolation is still forced by climate and the scale of distance to become part of a distribution system thousands of miles in extent. Indeed, one of the unpleasant characteristics of modern industrial life is that regional differences tend to be ironed flat by the economic pressures toward uniformity. The differences between small villages in separate countries are now far less than they were a generation ago, and something has been lost in that transition.

The communal enclaves of nineteenth century America, the Shakers, the Mennonites, the Pennsylvania Dutch, the Oneida Community, and others, nearly all consisted of groups each of which was united by an unvarying, agreed-on plan for how people should run their lives. Those who have lived in and then left the modern communes tell us that however the codes of behavior of these organizations may differ from the norm of the world outside them, internally they have rules strictly maintained. This should be no surprise. A commune is the limiting form of a small, isolated village, and as anyone who has lived in such a place can testify, social pressure there is almost always far stronger than in the anonymity of a large city.

In contrast, and very much by intent, I have said nothing about the government of space communities. There's a good reason for that: I have no desire to influence or direct in any way, even if I could, the social organization and the details of life in the communities. I have no prescription for social organization or governance, and would find it abhorrent to presume to define one. In my opinion there can be no "revealed truth" about social organization; there can only be, in any healthy situation, the options of diversity and of experimentation. Among the space communities almost surely there will be some in which restrictive governments will attempt to enforce isolation, just as such governments do on Earth. Others, hopefully the majority, will permit travel and communication. Within the brief time of twenty years, during which transatlantic air travel has gone from the unusual to the commonplace, we have seen how powerful a lever it has been for the transmission of experience from one country to another, especially among the fraternity of young people. Logically, if the cost of transportation between the communities becomes as low as it is now projected to be, travel between most of the communities of space will be far more frequent than it is now between nations on Earth, and people will be able to form their own opinions, on the basis of direct observation, as to what constitutes successful or unsuccessful experiments in government. With energy free to all, materials available in great abundance, and mobility throughout the solar system available to an individual community, it should be more difficult in space than it is on Earth for an unsuccessful government to argue that its failure is due to unavoidable circumstances of location or resources.

There is another profound difference between the historical Utopian attempts and the humanization of space. The communities of the past were formed on the basis of new social constructs, but took their technology from the world around them. Some even made a conscious selection of more primitive or more restricted technological equipment than available in the world outside. In extreme form this tendency shows in the prohibition, among several of the existing Utopian sects, of any equipment for day-to-day living more advanced than that which was available in the nineteenth century.

The reason for this restriction, usually clearly stated and understood, is the need to prevent "contamination" of the Utopian social ethic by contact with the outside world. There is recognition by the leaders of the enclave that its social organization is unstable, and can only be maintained by isolation. Usually, the "danger" to the maintenance of that unstable situation is that young people from within the enclave will learn of the additional choices available in the world outside, and will insist on leaving to enjoy them.

I share with many an admiration for the Utopian groups that have managed to retain their identity and values through several generations of rapid change. Those of us who might have been tempted, during the decade of the 1950's, to feel concern and even sorrow because of the narrowed horizons permitted to the children of such groups surely felt quite differently during the 1960's, seeing an epidemic of drugs and a lack of purpose spread throughout a generation in the world outside. It may even be that

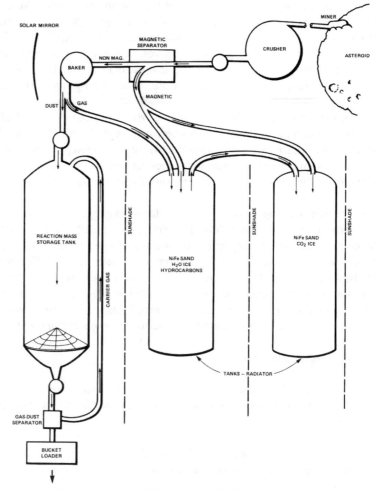

Processing of asteroidal materials will utilize solar energy and solar heating (concentrated by mirrors), so that converting the raw materials into useful elements can be done enroute to L5.

among the existing Utopian groups there are some free of anti-technological taboos, who will find it easier to retain identity by resettlement in space than to remain on Earth. The humanization of space is though no Utopian scheme. The contrast is between rigid social ideas and restricted technology, on the part of the Utopias and communes, and the opening of new social possibilities to be determined by the inhabitants, with the help of a basically new technical methodology, on the part of the space communities.

One can speculate, with some supporting evidence, that as a result of the individual choices that led to the historical colonization movements on Earth, there are now subtle but real differences in attitude toward change and further migration on the part of the people in the old and the new countries. Here in the United States, and in Canada, Alaska, Australia, and other former colonies, there may be a greater restlessness, a greater desire for travel and change, than exists in those populations descended from the people who stayed at home rather than emigrate. Of the thousands of letters I have received about the space community concept, a disproportionate number come from the lands that were once colonies. Already, from the many letters that express a personal desire on the part of the writers not just to support but to take part in the outward venture, it is clear that the early settlers in space will be exciting people — restless, inquiring and independent. And quite possibly more hard-driving and possessed by more "creative discontent" than their kin in the Old World.

In space, where free solar energy and optimum farming conditions will be available to every community, no matter how small, it will be possible for special interest groups to "do their own thing" and build small worlds of their own, independent of the rest of the human population. We can imagine a community of as few as some hundreds of people, sharing a passion for a novel system of government, or for music or for one of the visual areas, or for a less esoteric interest: nudism, water sports, or skiing. Of the serious experiments in society building, some will surely be failures. Others, though, may succeed, and those independent social laboratories may teach us more about how people can best live together than we can ever learn on Earth, where high technology must go hand-in-hand with the rigidity of large scale human groupings.

Just as happened during the settlement of the American West and of Alaska, when the population at L5 increases in number some of the pioneers may be the sort of people who will say: "It's getting too crowded around here; let's move on." Those people may be among the most interesting and productive individuals. They may want a more complete independence, and so may decide to go homesteading just as did our great-grandparents in the mid-nineteenth century American plains states.

Here, now, is one way in which a pioneer family might go about a homesteading venture. Though the details will surely be different from those I describe, each possibility that I will give is based on a number that can be calculated, or on analogy to similar situations here on Earth. I am giving it in the form of excerpts from a diary, written perhaps in the early years of the next century. That too is by analogy. One of the relics of my family, preserved through five generations, is a book by an old lady who must have been, in her time, a holy terror. In her eighties she wrote an account in verse of a time when she had traveled with her seven sons across the plains of America in a covered wagon. In their travels the little band encountered dangers that space settlers will not face; hostile Indians, snows, exposure and short rations.

*July 15, 20–: Dear Stephen:*

*Your Mom and I are going to write down a record of our trip, to go with the pictures we're taking. Then when you're old enough to read and be interested in it, you'll be able to see how you came to be a youngster living in the asteroid belt.*

*It's been five years, now, since I joined the Experimental Spacecraft Association. We have an active chapter of it here on Bernal Gamma, and several of the guys in it work with me in the construction business.*

*If we were back on the Earth now, and got any wild ideas about setting out on our own to travel in space, we'd be out of our minds. A spaceship that could lift its own weight, and go through the split second timing that you need for a lift-off from Earth, would be a lot more complicated and expensive than anything that home craftsman could build.*

*Out here, though, we're in much better shape to go voyaging on our own. Our spacecraft never has to take big forces, and our engine can be small, because we don't mind taking quite a while to get somewhere.*

*With what we'd saved, and the sale of our house on Gamma, we were able to start with about $100,000. For the past three years I've been working on the spacecraft, and we'll hang on to it when we arrive in the asteroids, so it'll still be around when you're old enough to remember things. The Lucky Lady is a sphere about three stories high, made of aluminum because that's easy to weld. I've been building it in the marina, near the docking ports of Gamma, and we've checked the welds with x-ray equipment that we've borrowed from the plant. Alongside the Lady, at the marina, there are four more of the same kind. Chuck and Bill and the others will be going with us, in a "wagon train" of five craft, so that if any of us runs into trouble either before or after we arrive, there'll be help near at hand. Between us five we've bought a complete spare engine and a lot of spare parts and one-of-a-kind tools. When we get to the asteroid belt we can team up for big jobs when we have to.*

*Our plans came out of Spacecraft and Pilot, and were checked over by astronautical engineers before they were published, so they're sound. The Lady has a triple pressure hull, each layer a millimeter thick, and any one of the layers would be enough to hold a lot more pressure than we'll need. Altogether the bare hull weighs about 3 tons, and there's a lot of my time that's gone into it. The marina doesn't rotate, so all the construction was done in zero gravity. That way, I could handle the big sections of aluminum by myself.*

*Around the hull there's a layer of sand about a foot thick, to protect us from some of the cosmic rays and from solar flares. Outside the sand is a fourth shell, of very thin aluminum just to hold the sand in place. For extra help in case of flares we've also got a "storm shelter" outside the sphere in the form of a small aluminum bubble connected to the big one. There, the shield is a lot thicker, and if a flare starts we can be in the storm shelter in less than a minute and can stay there for several days if we have to. Babies are extra sensitive to cosmic rays, so the "storm shelter" is your nursery too.*

*We bought our rocket motors new. They're from the same company that makes them for the small Coast Guard rescue boats. Each one gives a thrust about as much as my own weight, and a bigger chunk of our "grubstake" money went into those than anything else. I understand they cost about the same as a small jet engine on Earth. Our life support air recycling system was bought used, rebuilt and recertified by the Federation Astronautical Agency. It too came off one of the Coast Guard boats, and we got it cheap, but I know that the government paid a lot more. They've gone over to newer models now.*

*Back on Earth, before your Mom and I moved out here, I used to belong to an Aero Club and flew little airplanes for fun. Things happened fast there, and navigation in bad weather kept me on my toes. I'd have to keep track of the Omni-Range signals, and the direction finder sometimes, and stay legal as far as the control altitudes and the rest of the regulations went, and all the time fly the airplane by reference to the compass and the gyros. Going from here out to the asteroids I won't have to worry about all that — there's no weather in space, so we'll be able to see where we are and where we're going all the time. We'll have two systems for navigation. One of them is as old as sailing ships on Earth's oceans: it's a sextant, to measure the angles between the visible planets and the Sun. That would be enough to do the job, but we've also got something else. Nowadays there are big transmitters set up in the orbits of Earth and Mars, and they send out pulses so we can calculate our position just by a simple radio receiver. On Earth's oceans they used the same method for navigation and called it Loran. With the handbook of transmitter positions and times that we've got, we can figure our position to within less than a mile, even though we may be twenty million miles out.*

*We went a bit overboard on radios, and bought three, all alike. They're about the size of the ones used in small airplanes. We'll use them for voice communications between the five families traveling together, and for dot-dash Morse code to check in with the Coast Guard. We're going to be on a flight plan and will have to check in once every three days. To do that, I'll be aiming the big aluminum foil dish antenna that I've built, using a little telescope to point it exactly back to the location of the receiver at L5.*

*Aug. 1st, 20–: The Coast Guard and the FAA people have been aboard, and we've got our clearance. They checked our Space Worthiness Certificate (Category R, Experimental Homebuilt) and our radio licenses, and my pilot's license (Private Category, Deep Space Only, Flight Within Planetary Atmospheres Prohibited). We've got food on board for two years, if we have to stretch it, lots of seeds, fish, chickens, pigs and turkeys. To get things started when we arrive, we've sunk about half our grubstake into prefabricated spheres and cylinders, aluminized plastic for mirrors, chemicals for crop growing, and a lot of equipment.*

*Aug. 8th, 20–: The Lucky Lady, loaded, shielded, and ready to go, weighed in at close to 500 tons, so we didn't take off from Gamma with any big burst of acceleration. We weren't up even to walking speed a minute after we started thrusting, but our speed slowly built up and now, after a week, we've gone farther than the distance from the Moon to Earth. It'll be another eight months to go, about as long as your great-great-great-Grandad took to get from Illinois to California.*

*October 10, 20–: We've had a bit more excitement than we bargained for these past weeks. First of all, Bill's engine developed a problem. He wasn't getting the thrust that he should and the fuel was going too fast. Those engines are pretty complicated and we weren't able to solve the problem on our own quickly, so did an engine change to the spare. That wasn't too difficult — we just maneuvered the five spacecraft close together, docked them, closed up the hatch behind the engine, and did the engine change in our shirtsleeves. From now on we'll have plenty to keep us busy, because we have all the manuals on the engine and we're going to take our time and see if we can figure it out well enough to fix the one that we pulled from Bill's ship. While the engine change was going on we were "dead in the water" with no thrust for nearly four days, but here in space that doesn't mean an emergency. We still had our speed, and all that the lost time means is that we'll make a very small change in the thrust direction and take a little longer arriving.*

*Only two days after we got finished with the repairs, we got hit with our first big solar flare. Those things build up in minutes, so there wasn't time to get any warning. When the alarm bells sounded we all scooted for the storm shelters, and stayed holed up for three days. By then the flare had died down so much that our ordinary shielding was enough.*

*Nov. 23, 20–: We brought you out of the nursery so you could be with us for our Thanksgiving dinner: turkey, canned cranberries, and lots of extras we've been saving. So far we've got a lot to be thankful for. There were some colds early in the trip, but after that everyone's been healthy, and nobody's got any tooth problems yet. If we can last to the Belt, where there are dentists, we'll have escaped the biggest problem that hits groups like ours.*

*All of us have been using our time to get a head start on construction. We began with our assembly bay, and that's something the five families will share, 'til we can build more. It's a cylinder as big around as the Lucky Lady, and as long as a city block. It's made of aluminum sheets, and we made it without ever going out in "hard suits." We're in free flight now, the engines have been shut down, so we handled the construction bay by just clamping on to it with our grippers, very slowly walking the whole ship over to the place we wanted to work on, and then handling the welding equipment through sleeves that we've built in to each ship. I guess the setup is a bit like a chemist's dry box. The ends of the bay are hemispheres of aluminum, and when the last weld was done and checked the bay was a gas-tight chamber. We let the liquid oxygen tank get a bit of sunlight, so it would slowly boil off, and after a few days the oxygen pressure in the bay was breathable. We have all five ships locked on the bay now, so any of us can go in there to work, and that's where all the glasswork is being done. The welding, of course, is better done in vacuum.*

*Our first "dockyard" job has been the crop modules. Each one's a cylinder of a size that just barely fits in the assembly bay once all the pieces are welded together. When we're done we weld in a lightweight floor, and under that we set up the chicken coops and the pig pens. The roof is trickier, because we have to let in the sunlight. In the L5 communities they do that with thick metal webbing and then plates of glass to form the windows, but here we do things in an easier way. We have prefab aluminum sheeting that has a lot of small holes in it, and we seal over each hole a disc of glass with a plastic compound. When we finish a crop module we pump the oxygen from the bay into a cold storage liquid oxygen tank, and open the end-bolts and take off one of the hemisphere ends, and float out the finished section.*

*Dec. 25, 20–: You were out of the nursery again today, and all twenty three of us got together for a real big Christmas dinner. We had ham and a lot of frozen food, but next year, if we're lucky, we'll have fresh sweet potatoes and corn and fresh pumpkin pie as well. I've been whittling some new toys for you, and you seemed to go for them. I know you won't thank me for reminding you when you're a bit older, but Mom is proud that you say "Mom," and "Dad," and "ship" and "dog." I don't think Chuck's family would think of going anywhere without Snoopy, and if that other dog Maggie comes through like she looks, we're going to get one of the litter for you.*

*May 10, 20–: Looks as if we won't have time for any more writing for a while. We've been prospecting for the past month, and now it looks as if we've found us a good one. You couldn't even see it through a telescope from the Earth, but we figure it's got a mass of around three million tons — a lot more than we'll need even in your grandchildren's time. The little spectroscopes that we brought along tell us that it's got plenty of carbon (we picked the asteroid because it looked good and black) and there's nitrogen and hydrogen and plenty of metals too. So we've got some clearing and stump pulling to do, and by the time you're big enough to handle a welding machine you'll be my helper. We've got a whole world to build here, Stephen, so grow up fast and get in on the construction!*

The spirit of adventure, and the drive to be free and run one's own life, even at the expense of hardship, danger and deprivation, are as old as humanity, and must have been at the heart of the Westward movement as they will be for the migrations that will start at L5. If we traced the development of an embryonic settlement, of the kind that might begin with a trek of the sort just described, we might find that the pioneers would construct their habitats by the labor saving method of evaporation from an aluminum ingot suspended by magnetic forces in zero gravity, and heated by concentrated sunlight. Within two or three years a sphere with a land area of more than a hundred acres for habitation, and an additional several acres for crops, could be made in this way, most of it quite possibly by a housewife monitoring a control computer from her kitchen. A computer to do that job wouldn't be much more complicated than a pocket calculator, and a few decades from now a much more powerful computer installation, of the sort that's now found only in offices and laboratories, will be of desktop size and won't cost more than an automobile. Almost certainly each of the pioneer families will be equipped with one of them.

Examining growth rates, we find that the tiny asteroidal chunk described in the homesteader's diary would suffice for a population of at least 10,000 people, so there would be no need for the pioneer group to seek new materials for at least several hundred years, even if its population grew at the present world average rate.

In our modern world, with its concern for vanishing resources and for preservation, our immediate reaction on hearing of an available resource is to consider its protection. When I described the resources of the asteroidal belt to a group at the National Geographic Society, there was an immediate reaction: "Please don't take Geographos!" There need be no fear of that. Geographos is a small asteroid now thought to be of the stony-iron type, and should be safe from mining activity.

In the case of a growing technological civilization, with each new material resource we must associate a time scale. For example, if the total reserves of material to be found in a new "mine" will last only ten years, but if the new technology required to exploit that resource will take twenty five years to develop, the expected returns are hardly sufficient to justify the effort. Earlier I pointed out that the material reserves in the asteroid belt are sufficient to permit the construction of new land area totaling 3,000 times that of the Earth. In making that statement my purpose was not to encourage a corresponding growth of the total human population, but rather to suggest that materials limits alone should not be used as the justification for the imposition of limits on individual human freedoms. The freedom to have as many children as a family wants is by no means as important as the freedoms of speech, communications, travel, choice of employment, and the right to an education, but it is very difficult to abrogate one freedom without compromising others. As Heilbroner has pointed out, in a society held by law to a steady state condition, freedom of thought and of inquiry would be dangerous, and would probably be suppressed.[6]

In the same spirit, not of encouraging thoughtless growth but of opening possibilities which will encourage the extension rather than the curtailment of freedom, we can look beyond the materials limits of the asteroid belt and inquire as to the total resources of the solar system. I've argued that a growth rate of about a tenth as large as our present explosive increase would be sufficient to make the difference between stasis and change — it's just enough to be noticeable over the lifetime of a single human being. In the space communities, that growth could be matched by a corresponding increase in the total land area, rather than by additional crowding as on Earth. For that moderate rate of growth, the resources of the asteroids would be sufficient for at least four thousand years, at a population density the same as that of Earth (averaged over all the land area of our planet, including the desert, polar and wasteland areas now uninhabitable).

If we look beyond the resources of the asteroids, there are three further aggregations of materials within the solar system, each of which has a large total quantity: the moons of the outer planets, the cometary debris, and the outer planets themselves. As far as we know, all of these aggregations are without intelligent life, and all but the outer planets are invisible to us without telescopic aid.

The moons of the outer planets have a total quantity of material roughly 10,000 times that of the asteroids; the outer planets themselves, a thousand times more. The existence of those resources, beyond those of the asteroid belt, means therefore that even without the cometary material there would be enough for expansion at a moderate rate for more than twelve thousand years. Each of the new classes of material resource would permit, by its exploitation, several thousand more years of expansion, and the technology required for the opening of each resource would hardly require more than some tens of years to develop. Although I don't advocate it, I must conclude therefore that there is room for growth at a moderate rate for many thousands of years, should that be desired in every era by the human population then alive. Although twelve thousand years is short on the time scale of evolution, it's a very long time on the scale of social institutions. If we consider a voyage in time as far into the past as we can now contemplate into the future, we would be close to the time of the last Ice Age, long before the earliest beginnings of recorded history.

If long term growth may indeed take place, it is tempting to consider the corresponding increase in what we might call "capability," a measure of the power of humanity over the physical environment. We can only guess, but if we take the capability to be something akin to a gross national product, we may guess that it could be proportional to the growth factor itself (that is, to the crude ratio of populations), and to the productivity (the output per individual human being of some measurable product, either material or informational). If the latter is taken to be as little as 1.5 percent per year, and the former is 0.2 percent per year, the increase in total capability over so long a time as 12,000 years would be a truly astronomical factor of ten to the eighty eighth power. The implications of that increase in capability, admittedly speculative in the extreme, are fascinating to contemplate. Almost certainly they would include an enormous degree of control over the environment by each individual human being. Ten to the eighty eighth power, for example, is more than the number of the individual atoms in all the stars, planets, and dust clouds of our galaxy.

Evidently then, it's possible in principle for a civilization to advance from prehistory to a state of enormous capability on a time scale which is very short in galactic terms — less than one part in 200,000 of the age of the Sun. Why, then, has no previous "explosion" of a civilization into a situation of great physical power not left its mark on the galaxy? Why are there no beacons burning to light our way? Perhaps the birth of a civilization capable of migration into space is extraordinarily unlikely, or perhaps social instability and stagnation are overwhelmingly powerful civilization killing forces, or perhaps — as I have suggested earlier — moderation and empathy come with technical maturity, and there do exist long lived galactic civilizations, all of which prefer, for our own good, to let us develop on our own.

# Chapter 12
# The Human Prospect In Space

Speculation about a development still in the future is a scary process from the viewpoint of a scientist. He is used to making predictions which cannot be proved or disproved until later, but he makes them on the basis of experiments, carried out with all the care and diligence that he can muster. If he has maintained a sufficiently high professional standard in his experimental technique, he knows that later work can only prove him right. When a scientist indulges in speculation, he throws away the experimental tools which give him his only claim to authority and expertise, and his predictions do not deserve much more weight than those of anyone else. Even so, inevitably I must indulge now in speculation, and I do so with considerable apprehension, knowing full well that I am edging out farther and farther on a very shaky limb. Like an automobile driver in winter on an icy road, trying always to keep at least one pair of wheels on the solid pavement, I will try to keep each speculation within the bounds of numbers which can be calculated.

History and analogy are solid ground within the treacherous marsh of speculation. We know that foreign trade has been the economic basis for most of the successful human colonies in their early stages. For the long term economic viability of communities in space, we expect that there must be something which Earth must buy from L5, and something that the residents of L5 must import from Earth.

The need for cheap energy on the surface of Earth, in the form of electricity transmitted by microwave from solar power stations in orbit, is likely to exist for a long time. Even if per capita income in the developed world remains frozen at some level, for several more generations there will be a demand for more energy every year, as the Third World struggles to achieve economic freedom and take its place in the community of nations. While that demand continues the L5 communities should find a ready market.

The suitability of L5 as a location for the production and use of heavy scientific equipment (telescopes, research spacecraft both manned and unmanned, and laboratories for the study of zero gravity conditions) should give the residents of L5 another sector of trade with the inhabitants of the Earth.

In my view, the likelihood that marketable products can be returned profitably to the surface of Earth from L5 is much more doubtful. That return would require throwing away the single biggest economic advantage that the L5 communities will have — their location at the top of the 4,000 mile high gravitational mountain which towers above us here on Earth. Nevertheless, some consideration should be given to this possibility. The mechanics of payload return have been considered by Eric Drexler of M.I.T. He concluded that shipments of materials from L5 to Earth might best be made within reentry bodies fabricated of titanium. The plan would be to recover the lifting bodies from the ocean and break them up for the (high value) pure titanium they would contain. There may be a time when the economics of that process will be favorable, but I would be reluctant to invest in a titanium import firm myself.

The "products" needed at L5, and available from Earth, will change as the communities develop from one bare outpost to a thriving, booming frontier settlement. In the beginning, L5 will need machines, tools, computers and almost every other piece of complex equipment both for productivity and for life support. Carbon, nitrogen and hydrogen from Earth will be needed until the time when the asteroids can be mined.

We should recall the fact that the velocity intervals to L5 from Earth and from the asteroids are nearly equal. For that reason transport costs from Earth and from the asteroids may be comparable for a time. There may be then a period in which economic competition will tend to drive down freight rates for carbon, nitrogen and hydrogen, both from the asteroids and from Earth, although eventually transport from the asteroids should prove cheaper.

For a period of many decades during which the initial beachhead in space is expanding toward a mature community, L5 will need people, at a rate far more rapid than natural reproduction could supply. During all that time, the L5 communities will need to bring people from Earth, and we can expect to see, as we have in the case of Australia, a period in which free passage, initial personal "grubstake" capital, and perhaps initial free housing will be offered by the L5 communities as inducements to attract new immigrants from Earth.

The existence of those several components of a two way trade, in which both sides would benefit, should help to maintain a peaceful relationship between the L5 habitats and the nations of Earth. If irritations and misunderstandings do appear, as is inevitable in human relationships, fortunately neither side will be likely to risk a complete breakdown of trade. The price of serious conflict will almost surely be too high.

Though some items may be traded only for a short period, through much of the next century the need for additional energy here on Earth probably will ensure that L5 will have markets for new satellite generator capacity, and the communities' need for immigrants will probably continue about equally as long.

Ultimately, if the L5 civilization nears maturity and the Earth's population is stabilized, we can expect, in analogy with similar situations on Earth, that a two way tourist trade will become an important part of the economic picture. We can be almost certain that such will be the case when we realize that in each passing decade the cost of transport, in constant dollars, will decrease as technology advances.

It has been said that new wealth requires three components: energy, materials and intelligence. At L5 the source of materials will be inexhaustible at least for several centuries, and the source of energy will be reliable and effectively limitless for the next several billion years, to the best of our present knowledge. The third component is the human organization of machinery and human effort in a productive way. Productivity can be described by the ratio of output products to the input of human labor. If measured in non-monetary terms (tons per person per year) the ratio automatically takes into account the effects of inflation.

For many centuries productivity was static, held down by the limitations of a hand tool technology and the energy resources of human and animal labor. Then, with the industrial revolution, productivity began to increase. In the modern industrial societies of North America and Western Europe, that increase has averaged between 2 and 3 percent per year. (It's been argued that in a pure capitalist economy, without government regulation, the interest rate on capital should be set at the same figure. Inflation, now several times higher than the productivity increase rate, adds to both productivity and interest rates in a way that tends to conceal the underlying real changes.)

Individual wealth is proportional to the productivity if government does not absorb a greater fraction of the total wealth as time goes on. A productivity growth rate of 2½ percent is enough to double real (uninflated) per capita income within less than thirty years. As we view the goods and services available and normally used in the western world today by people a generation younger than ourselves, we see that indeed our area has experienced at least a doubling of the real income in a time of just one generation. In space, although not on Earth, it is conceivable that such an increase in productivity could continue for a very long time. In the U.S., at present, the annual family income is near $15,000. On Earth the limits on energy and materials are already putting the brakes on the increase of per capita income, but in space we can anticipate that by the year 2100, at a continuing growth rate of 2.5 percent per year, the average family income could reach the equivalent of more than $300,000 per year in non-inflated 1975 U.S. dollars. Logically, that increase can only occur if available energy also increases, to a total of about two hundred kilowatts per person in a space society of the year 2100. Some of the amenities which we might consider for the end of the next century will not be energy- or materials-intensive. Perhaps the outstanding example of a sophisticated, energy saving amenity is the electronic computer. Almost certainly by 2100 computers will reach a level of capability so high that nearly every every common, predictable task will be computer controlled, and will be carried out by machinery which in its turn will have been constructed in factories requiring very little human intervention. Other amenities will not be so economical of energy. Long distance transport, for example, even in space will require a certain amount of energy. Logically, we can expect that by 2100 ordinary people living in space will take for granted the availability of inexpensive transportation, energy intensive, which will give them tremendous freedom of movement over great distances at speeds of several thousand miles per hour. A two dimensional array of space communities large enough to house the equivalent of Earth's present population, each person having at his service two hundred kilowatts of electrical energy derived from solar heat, would extend over less than 3,000 kilometers. Given enough energy, in space a normal cruise speed of 3,000 kilometers per hour would be quite practical, for an engineless vehicle accelerated by an electric motor. The equivalent of a whole world in diversity of population, climate and landscape would therefore be available to a resident of space in the year 2100 within a travel time of an hour or less.

As the real income of the settlers in space increases, it seems unlikely that the residents of L5 will choose to remain in the rather cramped surroundings of the early habitats. On Earth, we are accustomed to the idea that with every passing year open space is enclosed, shopping centers spring up on once green meadows, and the pressure on wilderness areas increases. At L5, where the rate of construction of new land areas will be limited only by productivity, we can expect that over the course of a century the population density can go down rather than up, whatever may be the absolute population size and its rate of increase. We can estimate roughly the population density of a new space habitat built in the year 2100 by taking present day figures for the productivity increase rate and the world population growth rate. (Let's hope that's an overestimate: if it is the answer will be better than we're now calculating.) Taking the present U.S. value for the fraction of the population that's employed (around 40 percent) and assuming that a quarter of the work force is employed in the construction of new habitats, we find that each new colonist of the year 2100 could have an "allotment" of almost two thousand tons of structure.

To see how far this would go, we need a model. Island Two will serve — each such Bernal Sphere would have a structural mass of several million tons. Putting the numbers together, each Island Two, with almost seven square kilometers of living area, would be occupied by only one small village of twenty six hundred people. Country living indeed! In space, all agriculture and industry would be located in additional area outside the living habitats, of course, so the L5 land area would be fully available for living space, recreation, and regions allowed deliberately to run wild (much of what we now call "wilderness recreation" area here on Earth was logged off or farmed less than a century ago, so the notion of a deliberate wilderness should not be strange to us). Even before correction for agricultural and industrial areas, the density would be comparable to those of some of the countries in Western Europe (the Netherlands have one thousand people per square mile, and Italy four hundred and fifty, with all its agricultural, industrial and mountain areas included in the ratio).

The Bernal sphere external view and functional diagram. Well within today's technological capabilities, Island One offers much more than just a home in the sky for 10,000 people - it's a hope for the future for all mankind.

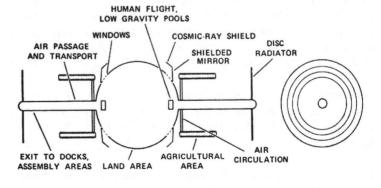

On Earth, even with the assumed success of population control programs, the total population will rise to at least ten billion some time in the next century. On the average we should assume then that population densities here will just about triple, until substantial emigration to the space communities takes place. Crowding, already severe in some areas on Earth, can be expected to get worse. In contrast, if we follow the population density projection for L5 another century into the future, we find for the space habitats a density less than a third of that now in mountainous, pastoral Switzerland, and considerably less than the average that the whole Earth will have by the early 1990's.

With increasing automation, it seems likely that the "standardized" portions of a new habitat — outer shell, mirrors, shielding, heat radiators and other externals — eventually will be constructed almost entirely by automatic machinery. Human intervention will be needed in just those areas where creativity and imagination will be called for: landscaping, architecture, and perhaps new artistic specialties like weather design and creative ecology. It may be that a group of settlers taking possession of a new land area in a habitat built by machinery will prefer to do the finishing operations themselves; to add the human touch by landscape, architecture, and choice of plant and animal species. Their first years could be spent in a way similar to those of our pioneer ancestors; each passing year would bring a sense of accomplishment and the pride of putting an individual stamp on home, garden and forest.

Specialists argue about the reasons for inflation. Even now, after many decades of effort and study, economists are not in full agreement about its causes. The simplest of all explanations is still in as good favor as any of the more complicated: that inflation is caused by "more and more demand chasing fewer products." A number of the factors which may drive that supply / demand spiral on Earth would not be present, or would be much reduced, in the space environment. As noted before, energy costs at L5 can be expected to decrease continually with the passage of time, because the source will be free and limitless, and technological advances can only serve to improve the efficiency with which that solar energy is converted to usable forms. Once the asteroids become reachable for mining, every chemical element will be available in abundance, and the solar driven transport systems for returning those elements to their points of use can only improve in efficiency and decrease in cost as technical development continues.

Here on Earth there is an inflationary pressure of the classical "more chasing less" variety, which we can observe in action every day. As the population density increases, land prices are driven up inexorably. Each time a new housing development is opened, prices are at a minimum. Then, as the number of vacant lots decreases, prices go up until the last few go at prices almost of the sellers' choosing. If one wants to see inflated land prices, one need only look at desirable places where zoning laws keep the number of new building lots strictly limited, and where there are plenty of rich buyers searching for land. Switzerland is an outstanding example.

In the space communities population densities should decrease rather than increase. There should be no shortages of energy or materials. Perhaps then in the space environment there will exist the best conditions for a non-inflating economy. If, over a period of many decades, severe inflation continues even in space, then our descendants must conclude that the main causes of inflation are not material but psychological. Even in that area the space communities may be at an advantage. We know that a primary psychological reason for inflation is fear — fear that some necessity or a product not essential but highly desired is going to "run out" so that an unreasonable price for it is justified; the "stockpiling" syndrome. Under the conditions of the space communities, after the first decades of learning and growth, it will be relatively difficult to create in the minds of the settlers the conviction that something material will soon be in short supply.

More uncertain than almost any other prediction about the future is any statement on the long term effects of the space environment on the length of human life. Even so, we can make a plausible case for the statement that human life will be extended in space, though it will take some time before the prediction can be checked.

First, the fundamental conditions for the maintenance of life should be at least as favorable in space as the average in desirable areas on Earth, and far more favorable than in the regions in which most people now live. Poverty is a killer, and the wealth of space should permit most of the total human population to escape from poverty. Atmosphere, temperature and sunshine in space can be optimized for good health. Given the shielding that can be obtained by the proper use of industrial slag, the radiation intensity in a space habitat should be no higher than it is on Earth. The risk of accidental death should be lower rather than higher in space. What though of the elderly? Here on Earth, with age and the infirmities of age, the body must spend more and more of its reserves of energy in simply fighting gravity. In the institutions to which many elderly people migrate, a great deal of the equipment one sees is devoted to the single task of assisting the body in its eternal battle with gravity.

In contrast, we can imagine that in a space habitat anyone with difficulty in walking will spend most of his time at a high elevation where gravity will be reduced. Those who would be confined to bed on Earth could have freedom of movement in a region of near zero gravity.

Cardiovascular ailments are among the major causes of death for the elderly. In space we can expect that people with problems of circulation can move to low gravity regions, and there enjoy freedom of movement and moderate, non-tiring exercise. In summary, it seems quite possible that people in a space habitat will live to a greater age than they would on Earth. Perhaps it is even more important that their later years could be spent in conditions of far greater freedom and independence than their physical condition would permit them on our planet.

In the earliest of the technical notes on the modern development of the humanization of space I commented on the possibility of reducing the population of Earth by migration, perhaps by the middle years of the next century.[2] In doing so I emphasized, as I always must when the topic is raised, the difference between possibility and prophecy. If human migration into space does occur, it will certainly have the power eventually to permit such emigration, as you can prove with even the simplest of pocket calculators, using the numbers I've given in the last few pages for the mass of an Island Two habitat, and taking its population as 140,000. You need one more input: the fraction of the labor force engaged in habitat construction; let's take that as half. We're being a bit conservative by assuming an Island Two geometry. Island One weighs less than half as much per person, as far as structure is concerned. Even without allowing for any productivity increase over the twenty five tons per year that's now common in heavy industry on the Earth, you'll find a doubling time of only seven years for new land area in space.

I've assumed that the present day value of productivity will still be appropriate to the year when the first Island One is finished. We'll take that as our "time zero." There might be a "dedicated" period of intensive construction after that time, as many of the nations of the world hasten to gain a foothold in space. In that pioneering era most of the space dwellers (perhaps four fifths) might be employed, and two fifths of all their products might be new habitats rather than such things as satellite power stations. In that case, the doubling time for land area in space would be only two years, and in just eight years there might be 160,000 people living in space.

Let's trace what would happen if then the employed fraction dropped to the U.S. average, the colonists switched to building the larger Island Two habitats, productivity continued its present slow rise and — just as an exercise — all the output productivity in space went into new habitat construction. How fast then could the population in space grow? (Notice I'm saying could, not will.) Again it's easy with even the smallest pocket calculator, and the answers are:

| YEAR | POPULATION |
|------|------------|
| 10 | 290,000 |
| 15 | 1.5 million |
| 20 | 9.2 million |
| 25 | 68 million |
| 30 | 631 million |
| 35 | 7.3 billion |

Before challenging these numbers, note that they're based on a continuation of the present slow rate of productivity increase. Without that, the time scale would be somewhat longer, though not much. The population shown in the table for year 30, for example, would be reached about five years later.

The point of this calculation is that the productivity that we have achieved already on Earth, when employed in the energy rich, materials rich environment of space, could lead within less than two generations to a production rate of new land area great enough even to accommodate the population increase rate of Earth. If the number of people on our planet rises to ten billion, and if its rate of increase goes unchecked, that rate of increase will be 200 million people per year. In the table, it would require only thirty years from the completion date of the first community before new lands would be increasing more than fast enough even to cope with such demands.

That exercise is not presented as an "optimum scenario." Indeed, I would much prefer to see our growth rate here on Earth decrease with time. But I would like to see it decrease for the right reasons — security, a decent standard of living, and free choice — not for what seem to me the wrong reasons: legal or economic coercion.

The second part of what one might call the "emigration problem" is transport — is it reasonable to consider a transportation system with the capacity to cope with such rates? Again surprisingly, the answer seems to be yes. In Chapter 10 I described a relatively near term vehicle system based only on the technology we believe we now understand. The fleet of vehicles I described would be capable of carrying about five hundred thousand people in one year from the Earth to L5. In the "fastest possible buildup," that emigration rate would be reached in about year 15 from the beginning of the Island Two era and a rate of two hundred million per year would be reached about fifteen years later.

To accommodate that higher rate, we'd like to have power supplies for shipboard use with a mass just under a ton per megawatt. That could come about either by several decades of development of solar cell technology, or by the use of microwave or laser transmission of power in space. With such performance the round trip travel time for a large ship powered by a mass driver engine could be as little as twelve days, with the outbound trip taking only three and a half days — less time than is required by the fastest ocean liner for an Atlantic crossing. If each ship were to carry 6,000 passengers, a modest increase in capacity over a fifteen year period from the time of the *Tsiolkowsky* and the *Goddard*, then about eleven hundred ships in all would be needed. That's comparable to the number of large ocean vessels that now sail the waters of Earth. If we check the productivity required for the construction of eleven hundred large spacecraft we find that their total mass would be some ten million "deadweight" tons, and that they could be built in three years by a work force of fewer than 0.1 percent of the population that L5 would have in year 25.

Transport from Earth to low Earth orbit, during the same era, would presumably occur in vehicles with passenger cabins as large as those of a Boeing 747. Compared with the capabilities of the present space shuttle, that's an increase over a time of around fifty years that's much more modest than our own experience in aircraft — from the 24 passenger DC-3 to the 400 passenger 747 in only thirty years. The trip from Earth to low orbit would take less than half an hour, whatever the vehicle size, and for a round trip time of four hours the transport demand could be met by a fleet of less than two hundred vehicles. That's only a tiny fraction of the number of large aircraft (about four thousand) already in the world's commercial jet fleet.

Ticket costs calculated by the same methods used earlier would be about $4,500 per person in today's dollars; comparable to the present cost of a round-the-world trip, and equivalent to only a few months' earnings under the conditions prevailing in the communities.[1]

From the industrial societies of North America, we pour into the atmosphere each year about ten tons of combustion products for each member of the population. Over a lifetime each person is therefore accountable for more than six hundred tons of combustion gases and smoke. By contrast, the fuel used to lift an emigrant to low orbit from the surface of Earth, by vehicles no more advanced than those of the present day, would be less than three tons — only as much as he would be using in a four month period on Earth. Once an emigrant left, the corresponding burden of his energy usage on Earth's resources and atmosphere would be lifted, permanently except for his later visits to the home planet. If the traffic to and from space ever reaches the frequency given in the example it will be very important to design engines for clean burning fuels, and to pay special attention to the delicacy of the atmosphere's ozone layer. There will be at least forty years of time to study the problem before it will be necessary to solve it, so I think we can conclude that there are no serious obstacles to handling even as great a traffic volume as has just been calculated.

When we consider this possibility of reducing the population of Earth by emigration, it's important to distinguish possibility from prophecy. As we have seen, the combination of technique with natural growth in capability would have the power to permit such emigration. Whether or not large scale emigration will occur will depend on how badly it is needed, and on how attractive the space communities become. With four billion people, Earth is already overcrowded in many areas; many would choose to flee Earth if it had ten to fifteen billion.

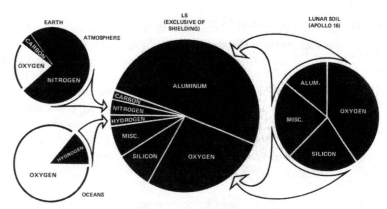

Combining where each element is most abundant with which sources are the most economically viable and ecologically sensible, we have a clear resource blueprint for the development of space. Recent knowledge gained about near-Earth asteroids make the situation even better.

The availability in the space habitats of high paying jobs, of good living conditions, and of better opportunities for children may stimulate the emigration of a considerable segment of Earth's population even if overcrowding on Earth is less serious than now appears likely. In the long run, because of the availability in space of unlimited cheap energy, of abundant materials, and of efficient combinations of attractive living area with nearby industry, I suspect that Earth based industry will be unable to compete economically with space based industry. If so then, as has occurred many times in Earth's history, people will follow the availability of jobs, and that will mean emigration.

A non-industrial Earth with a population of perhaps one billion people could be far more beautiful than it is now. Tourism from space could be a major industry, and would serve as a strong incentive to enlarge existing parks, create new ones, and restore historical sights. The tourists, coming from a nearly pollution free environment, would be rather intolerant of Earth's dirt and noise, and that too would encourage cleaning up the remaining sources of pollutants here. Similar forces have had a strong beneficial effect on tourist centers in Europe and the United States during the past twenty years. The vision of an industry free, pastoral Earth, with many of its spectacular scenic areas reverting to wilderness, with bird and animal populations increasing in number, and with a relatively small, affluent human population, is far more attractive to me than the alternative of a rigidly controlled world whose people tread precariously the narrow path of a steady state society. If the humanization of space occurs, this vision could be made real.

Science fiction writers are fond of assuming conveniences such as faster-than-light travel ("Warp Factor Six, Mr. Sulu"), suspended animation and teleportation. When speculation is involved I find it more challenging to see just how far we could go *without* assuming any science beyond that of our own time.

Earlier I described an asteroid voyaging research vessel capable of roaming the inner solar system with a laboratory village of several hundred people. In space the limits on the size of a vehicle would be far more relaxed than they are on Earth, and there's no reason in principle for not considering much larger mobile objects. A habitat the size of Island One could be equipped with a solar electric propulsion system of the kind described in Chapter 11. Human populations of 10,000 have existed in isolation for periods of many generations, within the history of our planet, and that number is quite large enough to include men and women with a wide variety of skills. Space dwellers will be well equipped psychologically for distant voyages, and a few decades after the beginnings of the human settlement of space there may well be large groups of people roaming the outer reaches of our solar system, on long term missions with a scientific purpose. Such groups could be connected intimately to the rest of human society by television and radio, so there would be no reason for them to remain isolated unless they chose isolation for reasons of their own. Even at the distance of the planet Pluto, the most distant known member of our family of planets, new examples of the visual and musical arts could be received with a time delay of only a few hours.

We can estimate the approximate limits to which a roving space habitat could go by assuming 1) that its inhabitants would want full Earth normal sunlight, 2) that the total land area for habitation and for agriculture would be that of Island One, 3) that a solar electric generator supplying the habitat with the present U.S. total per capita energy usage would be a desirable accessory, and 4) that the mass of a collecting mirror to concentrate sunlight should not be more than double the mass of the habitat. The corresponding limit of distance, if the mirror averages several wavelengths of light in thickness, is roughly four light days — about ten times as far out as the orbit of Pluto. That limit, approximate rather than exact, corresponds to a kind of "continental shelf" for our solar system; beyond it lies the abyss of interstellar space. Within that limit, though, it seems that there is no reason why a roving community should have to endure conditions any less luxurious than those of the habitats nearer Earth.

A spacefaring laboratory would be a gossamer-like affair, with a huge paraboloidal mirror. At the center, spider-like, the shielded habitat itself would rest, absorbing the solar energy collected from several thousand square kilometers of area. Within, I suspect that the traveling laboratory would be landscaped in a manner expressive of the inhabitants' need for the psychological solace of lush vegetation. Long time inhabitants of such communities would probably develop a passion for gardening, not only for flowers but for unusual vegetables and spice plants.

Of the inhabitants, in that case necessarily limited to a constant population, about a quarter would be of school or college age; enough to require a small university. Half the population would be within the normal working years, and of those people half again might be needed for all the services of the community: teaching, agriculture, maintenance, engineering, navigation; people to operate stores, print shops, cinemas, hospitals, libraries and restaurants. The replacement of durable goods by up-to-date equipment manufactured according to plans radioed from L5 might occupy another fifth of the work force. The remainder, perhaps 2,000 people, could be directly engaged in research: planetary astronomy, geology, geophysics, interstellar astronomy, and the operation of long baseline radio telescopes in partnership with laboratories near Earth. A laboratory of that size would be comparable to the whole faculty and staff of a medium size university or a large national laboratory of the present day. It would be quite large enough to carry out, over a period of years, thorough and systematic explorations of the outer planets, sending down manned and unmanned probes to the planetary surfaces for short excursions.

Living in a community like that would be rather like living in a specialized university town, and we could expect a similar proliferation of drama clubs, orchestras, lecture series, team sports, flying clubs — and half finished books.

Guessing at the deep space activity within the limits of our solar system's "continental shelf" during the next century, I suspect it will be confined to asteroid mining, to communities roaming the solar system for research, and to small fixed research colonies on habitable planets. The bulk of the human activity, in my guess, would be concentrated in the region near Earth and in the asteroid belt, and would be linked by a communications network with delay times given by the speed of light, and therefore no longer than about half an hour.

Our first good look at nearby star systems probably will be through large, composite (many-mirrored) telescopes based at, but not on, space communities. Perhaps our descendants will find, some time in the next hundred years, that a star within a few light years of our own is sufficiently interesting to warrant a closer look. That might occur if the star were proved by telescopic observation to have planets, for example. In that case a robot probe could be dispatched on what would be a voyage of many years. The most economical way to gain information about another star system would be through a "fly-by." The probe would use all its store of energy and all its reaction mass simply in accelerating, in order to minimize the travel time. It would plunge through the target star system at a speed of perhaps a tenth that of light, and in a few hours would gather all the information that its sensing elements could retrieve. Then, over a time which might well be measured in still further years, it would radio back to its human masters all the information collected during its brief hours of intense activity. As we view the rapid development of computers and miniaturized electronics, it seems safe to state that a century from now a robot probe could be far more reliable and sophisticated than any possible human crew, so our first close look at another star system is most unlikely to be through human eyes.

Could a space community some day ever venture beyond our continental shelf and embark on a trip to another star? If the community were large enough to constitute a complete society, and if the social stability of an isolated large group turned out to be great enough, such a voyage certainly would not exceed the bounds of physical possibility. But for that we must carry our speculations well beyond the limits of present day technology. Vessels limited to engines that could be built relatively soon, within the next few years, and which would use solar energy to maintain Earth-like conditions, would be limited to distances of a few light days from the Sun. For interstellar distances a source of energy would have to be carried on board. Though the television series *Star Trek* assumes much technology contrary to physics as we now understand it, some of its technical paraphernalia make sense within the limits of our present knowledge. In particular, the "matter-antimatter pods" about which Engineer Scott is always so worried are quite reasonable, assuming the technology of a century or two from now. Particularly in space, without gravity to bother us, it would be possible to build up a quantity of antimatter. The cost in energy would be enormous, and at present our methods for producing antimatter are primitive and inefficient, but there is no reason why they should always remain so. The most convenient form in which to carry antimatter would be liquid or solid — frozen anti-hydrogen, at a temperature of only a few degrees above absolute zero, would be a good candidate. Its atoms would consist of antiprotons around which positrons would circle.

Returning to the example of the Island One community, equipped for distant voyaging, we can imagine its mass budget for a mirror being replaced by an equal quantity of frozen hydrogen and anti-hydrogen. The antimatter could be maintained, in the absence of gravity, by electrostatic fields requiring no direct physical contact. It could be protected from the ordinary matter cosmic radiation and from dust particles by a thick shield of ordinary matter, and in principle should last a very long time. When we calculate how long a space community could exist, in Earth normal energy conditions, on stored antimatter fuel, we find that it could last for several billion years! Certainly plenty of time for an interstellar voyage.

In the second chapter of this book I had occasion to quote the conclusion of Professor Richard Heilbroner regarding the outlook for humanity if its only ecological range remains on our planet. "If then, by the question 'Is there hope for man?' we ask whether it is possible to meet the challenges of the future without the payment of a fearful price, the answer must be: No, there is no such hope."

Those of us who have enjoyed since birth an adequacy or even an excess of material comforts are the first to describe them as of secondary importance. Not so the large fraction of the world's population that moves in pain and poverty from birth to the grave. As we view the problems which now face humanity as a whole, it seems inexcusable that now, late in the second century of the industrial revolution, that so much of humanity remains in need of even the essentials for health and a decent life. Clearly, given a worldwide dictatorship of unchallengeable military strength and egalitarian outlook, the wide disparities in the wealth of nations and individuals could be much reduced. By my standards, to accept such rule, even if there were any realistic way in which it could be brought about, would be indeed to pay "a fearful price."

As we look back on the times of which the human race is most proud, it's hard to escape the conclusion that they were times of diversity, competition, unpredictability and considerable confusion. We still recount in admiration and pride the philosophical, literary and dramatic accomplishments of the Greek city states. In that period were born many of our most cherished concepts of freedom and individual worth. Is it pure accident that the classical era was also a time of great diversity and of disparate, often conflicting ideas from small communities of no more than a few thousand individuals?

I wonder, too, at the age of darkness that followed the monolith that was Rome. There, if ever in the distant past, an organized state arose with the power very nearly to take over the entire world, with ideals and a concept of civilization not completely abhorrent even to our modern view. Yet that brief period of supranational organization was followed by many centuries during which, as we now see it, little of what we think of as "progress" occurred. Is there something in the concept of universal organization that is basically alien to humanity? Something against which the human spirit rebels? Perhaps so. The next period in which we as humans take great pride is the Renaissance and the age of exploration — certainly a time of great differences, great uncertainties, and unprecedented mobility.

As we view the next decades with the new option of the humanization of space, of one thing we can be sure: that those years will be unpredictable. New possibilities have appeared, and with them literally a new dimension in which humanity can move. From Flatlanders, we suddenly have the opportunity to become the inhabitants of a three dimensional solar system. Clearly our first task is to use the material wealth of space to solve the urgent problems we now face on Earth — to bring the poverty stricken segments of the world up to a decent living standard, without recourse to war or punitive action against those already in material comfort; to provide for a maturing civilization the basic energy vital to its survival.

These are the immediate problems, and I have attempted to show how these problems can be solved by ourselves. They don't require supermen with a more than human capacity for organization, cooperation and self-denial.

It may be argued that the exploration and the settlement of space is no more than a "technological fix" for problems that should be solved on a higher, more intellectual plane. Yet by our evolution we are closely tied to the material world. We are the descendants of the survivors, from many generations during which the maintenance of life was a struggle every day with the material world. Our history does not suggest that we are well suited to changing, overnight, to a species disinterested in material well being, with paramount concern for humanity as a whole rather than for a narrower group. Indeed, our loyalties are first to those few individuals to whom we are linked by close ties of genetic relationship. Only with an effort do we extend our concern to the town, the state, the nation and the world. As a species, we have solved our problems by technical means for millennia, and it would be surprising indeed if we could change our character so completely as to abandon the methods by which we have survived.

Earlier I contrasted the new ideas in this book with the philosophies of the classical Utopians. Will the space communities be free of conflict, free of misery, free of sadness? Certainly not, as long as they are human. Rather, we should hope that they will give added opportunity for that most elusive of human occupations, so fundamental as to be written into our Declaration of Independence: "The pursuit of happiness." Our country has not survived its first two centuries on the basis of promised happiness, but rather on the promise that the search for happiness could go on. I hope, and I think, that those people who have taken the concept of the humanization of space so much to their hearts have not done so in a misguided expectation of perfection. If their letters and their conversations are any guide, they appreciate how difficult will be the conditions and the challenges that they will meet in a new age of exploration and discovery. Yet even the opportunity to try new ideas and to break out in new directions is more than could be hoped for in a world forever limited to the confines of our planet. It is, after all, only a few thousand years, perhaps a hundred human lifetimes, since humankind first abandoned the nomadic existence of the hunter for the stability of farm and cottage. No wonder, then, that there should be, deeply rooted even after those years, a need within us to feel that boundaries can be broken and new paths explored.

What of the arts, and of letters, in a new period of expansion of the human spirit? Creativity is the most difficult of the human attributes to predict, but it's at least hopeful that the age of Columbus and of Drake was also that of Michelangelo and of Shakespeare.[3] At a more homey level, occupations that have a flavor of openness and nomadic existence have always been celebrated in our romances. In modern popular song the ever-moving truck driver has taken the place that was occupied a century ago by the cowboy. In the challenge of the first outpost in space, and of the voyages that may be undertaken by those who travel with their families to the lonely asteroids, there should be matter enough for song and story.

As we consider the human prospect in space, we know that where people are involved there will always be the potential both for good and for evil. Yet there seems good reason to believe that opening the door into space can improve the human condition on Earth. Relieved even a little from the drive to struggle with other nations for the diminishing resources of our planet, we may hope for a more peaceful future than will otherwise be our lot. Generosity toward the Third World, in its attempt to avert famine and to take its place among the community of nations, seems more likely to be shown if that generosity can derive from new, unlimited resources rather than from those we already find to be in short supply.

More important than material issues, I think there is reason to hope that the opening of a new, high frontier will challenge the best that is in us, that the new lands waiting to be built in space will give us new freedom to search for better governments, social systems, and ways of life, and that our children may thereby find a world richer in opportunity by our efforts during the decades ahead.

# Part Two
# The Time Is Now

# Chapter 13
# Space Robotics

by
David P. Gump
President, LunaCorp

The new era of space exploration will feature an unexpected abundance of robots doing tasks that weren't predicted.

The robotics population explosion is coming because robots are essentially mobile computers that benefit from wider trends in the computer industry — the swiftly rising capabilities, the plunging prices, and the ever faster Web connections. Every task a robot faces becomes easier and cheaper as the computer revolution rolls on.

Robots will be part of the Internet, able to call upon almost unlimited data banks to guide their actions. If a robot's pattern recognition system is stumped by a new object or sound, it will have all the online encyclopedias, magazines and newspapers of the world ready to assist.

Conversely, robots will transform the Internet by becoming its mobile eyes and ears. Robot borne sensor suites will expand the Internet's reach so that Web denizens will be able to take part in almost any human activity, including expeditions to the Moon and Mars. Our connection to the final frontier won't just be through a handful of camera toting journalists. We'll be able to flit among dozens if not thousands of live mobile robotic portals into the real work and excitement of the final frontier.

This future is probably attainable within a decade. We're close to a great leap forward beyond the dull automated probes now launched by NASA, because today's terrestrial robots actually have begun to evolve into machines that do remarkable things.

These new capabilities will be put to work in space thanks to a shift to private initiatives in space exploration. It's been technically possible to create more interesting and effective robots for many years, but NASA hasn't sent them to the Moon and Mars because "interesting" robots aren't needed to meet most scientific goals. For example, the Mars lander that crashed in 1999 didn't even carry a mobile robot. In fact, it had only one interesting new capability — a microphone to capture the first sounds from another planet. Yet this new capability was provided privately, by the Planetary Society, not NASA — and it was grafted onto a Russian part of the probe, not a NASA part. Future space missions, run entirely by the commercial sector, will deliver even greater public involvement, because business caters to the public's interests. And robots are central to the plan.

The second factor, the incredible shrinking computer, is a key to creating highly self-reliant mobile devices like planetary exploration robots. Enormous computing power is required for each of a robot's main functions, such as its machine vision system, its path planning algorithms, its voice recognition software and its continuous internal monitoring systems. For space robots with narrow bandwidth links to the Web, enormous local storage space is required to give a mobile robot the databases required to have even a minimum of what humans know as "common sense." Humans know that staring into the Sun is dangerous for our eyes, and that a steep rocky slope is harder to traverse than a gentle rock free incline. But rules like these have to be stored on board an exploration robot for instant recall, along with potentially millions of similar insights — including knowledge gained on the job, as it explores another planet. (Recalling this type of key survival data via the Web is too risky when the Internet link to Mars, for example, often requires a 40 minute round trip communications time.)

One key system for a perambulating robot is the ability to see where it's going, and pick a safe path to get there. The territory ahead can be mapped using radar, or by laser line scanners, but they have drawbacks including power consumption and possible hazard to any humans standing nearby. The Robotics Institute at Carnegie Mellon University has been a key institution in developing the stereo vision systems that closely mimic the human stereo vision system. Stereo pairs are compared by the computational system to determine where objects might block the robot's path. Then the path planning software kicks in so the robot can work its way independently to a goal set by its human owners, or it can work in "teleoperation" mode. Teleoperation means that a human is directing the robot second by second — but with its built-in intelligence provided by sensors and path planning software, the robot can compare the commands it is receiving with its knowledge about which routes are safe to take and which are dangerous. Exploration robots built by the Robotics Institute can refuse teleoperation commands that are risky — which makes it safe for amateur drivers to take control. The Nomad robot that traversed the Atacama desert in Chile in 1997 had this ability to follow Isaac Asimov's second law of robotics: a robot should protect itself from harm, except when this would conflict with protecting human life.

In the movies, futuristic robots never have any trouble racing through the real world with their human companions. Until recently, actual robots rolled about on tires and faced the same challenges as wheelchair users in navigating stairs or outdoor terrains. In 1994, the Robotics Institute demonstrated that robots can use legs to walk into a live volcano (Mt. Spurr in Alaska). The eight legs of the *Dante II* robot were split into two banks, and it moved somewhat crab-like over very unstable ground. The robot was able to sample gases at the bottom of the volcano's vent as boulders rained down from the caldera's steep slopes.

LunaCorp's Robot Nomad demonstrated the power and excitement of exploration via telepresence. This prototype lunar rover covered more than 200 km across the Atacama Desert in Northern Chile.

Honda shocked robot developers worldwide in late 1996 when it unveiled its human form robot, because its two legged machine was able to walk like a person, even keeping its balance as it climbed stairs. Honda later displayed an advanced prototype walker standing 1.6 meters high and weighing 130 kg that played soccer — poorly, of course, but the wonder is that it could kick a ball at all. (See details at www.honda.co.jp/english/technology/robot/index.html)

The Honda effort proves that one facet of a *Star Wars* C-3PO is possible — a humanoid form able to walk through a human environment with stairs, street curbs and the like. As computing power improves, it will become increasingly possible to create robots like C-3PO that converse with humans and carry out detailed reasoning and deduction.

Another striking example of sophisticated robots able to mix with humans is the Sony Corp. AIBO robot pet. When Sony introduced the ERS-110 AIBO dog in June 1999, the 3,000 units available in Japan sold out in 20 minutes (at $2,400 each) and the 2,000 units available in the United States sold out in four days. When another 10,000 units went on the market in November 1999, Sony received orders for 135,000 worldwide. The Sony experience proves that as capable mobile robots are put on the market, they will be snapped up by eager early adopters — providing the base to finance follow-on machines at even lower prices.

Today, programmers are just barely able to make computers understand what humans want from them, usually just in tightly limited circumstances. An example is the United Airlines reservations computer. You can get the status of incoming and outgoing flights by speaking — the names of cities, the time of day, and the like. The computing system knows that the possible words you could be using are limited, so it can make some pretty good guesses despite the wide variation in people's accents and speaking rhythms. Other voice recognition systems can take free-form dictation, but they have to be "trained" first in a particular speaker's nuances as he or she recites a known text to the system. Robots able to understand anyone's speech without training — like the multilingual C-3PO of *Star Wars* — are still in the future.

## The Earth-Moon-Mars-Asteroid Internet

Robots connected to the Web will be a key part of our space future when the Web comes to blanket the Moon and Mars, not just Earth. From whatever Web portal we may be using, we'll be able to link directly to robots as they explore space. Some of these robots will be solo explorers, boldly going where no machine has gone before. Others will be part of robot colonies, where diverse types of machines cooperate to build structures for a future human settlement, or to set up an automated observatory. Still others will be companions to human explorers and colonists, carrying out the manual labor for a settler or automatically documenting the samples collected by a scientist. Almost all of these machines will be online, allowing us to see through their eyes as the final frontier expands.

Why would a settler on the Moon allow someone on Earth to tap into his or her robot assistant? At least initially, the cost of off-world colonization will be so high that any extra revenue — such as from selling Web access to personal robots — likely will be crucial to economic survival.

In fact, many of the initial exploration robots probably will be financed primarily from selling this public access over the Web and via television. The LunaCorp plan for Moon exploration centers on this ability of robots to serve mass audiences in entirely new ways with high definition television and interactive Web experiences. The round trip communications time for Earth-Moon interaction takes at least 2.9 seconds, and potentially 4-5 seconds if the signal must be relayed from a receiving station on Earth via several satellite hops to a central command center. Although awkward, LunaCorp simulations have proved that human controllers on Earth can take this lag into account when directing a remote exploration robot.

This opens the possibility of a new era of space exploration where the humans initially stay home and transport their senses to the Moon to move about in the bodies of robots. This will greatly democratize space exploration — limitless numbers of people can link into the robots' bodies for a voyeuristic experience, and hundreds per day can take turns actually controlling each robot.

Because teleoperation of Moon robots from Earth is feasible, much of the Moon's initial human labor force may never leave the home planet. They'll commute to office parks like the rest of us, and strap into telepresence chambers that put their minds on the Moon while their fragile human bodies remain safely in Silicon Valley or similar environs. (See www.lunacorp.com for details.)

Telepresence also will upend the traditional preference people have for manned space missions over robotic ones. Today's manned missions are exciting with heavy television coverage, while antiseptic robotic probes send back a few still photos once in a while. In the 21st century, companies designing robot led missions will spend the money to have live digital television signals beamed back to Earth, instead of today's snapshots. And the robotic explorers will have personalities that will make them interesting guests for news interviews and talk shows.

On the Web, robots will actually provide an experience that's far superior to that of human space travelers. For example, a robot explorer will be able to carry out multiple chats with its fans on Earth — with perhaps 20 or 30 conversations going simultaneously. The robot explorer can be speaking in precise geological terms to scientific teams back on Earth at the same time it is waxing poetic for reporters, while it also speaks in very simple words for a grade school class. All the while, the complete record of the robot's exploits to date is available instantly to compare with what's being investigated at that moment. No human could provide that much interaction to the public.

The new era of robotic exploration thus will be one of mass participation via the Internet and digital television. Anyone with a Web connection will have greater access to the experience of space exploration than even the NASA mission controllers did during Apollo — with digital video, a chance to direct a robot's movements on another planet, and the ability to directly chat with those robots as they explore. Only people who actually travel into space will have a superior experience, and hopefully the robots' democratization of space access will spur a very large scale demand for this next step in our evolution into the solar system.

# Chapter 14
## *A Technology For A Better Future: Space Solar Power An Unlimited Energy Source*

by
Margo R. Deckard, M.S.
Space Frontier Foundation
Los Angeles, CA USA

There are 6 billion human beings inhabiting this living world. Six billion humans who place demands on this Earth. Humans who want the Western standard of living and who justifiably want all the conveniences of modern life. Since the standard of living of a nation is directly related to its per capita energy use, the world is filled with rapidly growing nations that need more energy. The U.S. Department of Energy projects that worldwide demand for energy will double in 20 years, and double again in the next 20. It's clear that current energy sources are going to be neither adequate nor environmentally clean enough to meet this need.

A fundamental challenge in the next century is how to meet the world's growing energy needs from an environmental perspective. We must meet this challenge to provide the opportunity for prosperity to all humans. Fortunately, the Sun supplies the Earth with an abundance of clean and natural energy. Space Solar Power, or SSP, is a means of collecting that energy and beaming it down to the Earth wherever it is needed. SSP may be the key to meeting this challenge. SSP could be an environmentally friendly, econo-mical energy producing tech-nology that simultaneously promotes the human realization that the Earth is an open system while protecting the Earth's fragile biosphere.

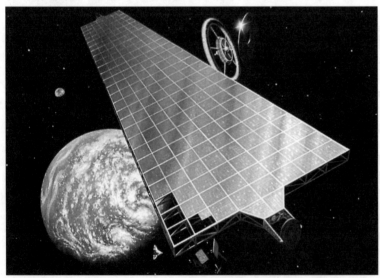

Satellite solar power stations are the key to providing unlimited power to both the Earth and to space colonies in a manner that will preserve, and hopefully improve, our ecology.

We must approach large technological solutions from an ecocentric point of view. Every day one hundred square miles of rain forest are lost. We are driving three species per hour to extinction. Undeniably, a suite of energy technologies will he necessary to meet the Earth's growing energy demands. Isn't it time that we invest in an unlimited energy source capable of supplying baseload power? An unlimited energy source designed with the environment in mind. There at least two communities that must be enrolled in the possibilities of SSP: the environmental community and the governments of the world.

The environmental community consists of academics, industry, advocacy groups and general enthusiasts. In 1963, Rachel Carson published *The Silent Spring* which pioneered a new environmentalism in which science would reach out to the public and educate them on the environmental impact of humans on the world. This increased public awareness has lead to "green" thinking. An ecologically minded public is demanding products that follow good stewardship practices. This market demand has led to green engineering practices which consider the entire impact of a product from manufacturing through to disposal.

Currently NASA, lead by the Marshall Space Flight Center, is engaged in studies to develop SSP. The goal of these studies is to develop a technology road map for the economical and safe development of SSP. The key to this road map is the development of commercial applications in the intermediate steps. This includes beamed energy utilization in space and improvements in solar cell technology. In conjunction with a technology development road map, a road map for the ecological design of this space based energy source must be developed. By involving the environmental community at this stage, we create a multi-disciplinary team to limit the impact of SSP while taking advantage of the environmental benefits of this technology.

Understanding the environmental impact is a surmountable challenge for the development of this technology. In the thirty years

This early test version of a receiving antenna for microwave (solar) power transmission will be replaced by a version that is transparent to sunlight but not microwaves, so that it could be located above ground level over safe grazing grounds.

that this idea has been studied, only limited data on the potential environmental effects has become available. There are many environmental issues which need to be studied including the long term effect on plant life under rectennas, the effect on telecommunications, the effect on objects flying through the beam, including birds, insects and aircraft, etc. Perhaps the largest environmental impact will not be from system implementation, but from the construction of the solar panel photovoltaics. While this industry is highly regulated, it still produces toxic waste products. Perhaps an answer to this problem is the production of photovoltaics on the Moon. This would definitely decrease the impact of system construction on the Earth and may decrease construction costs.

While these issues may seem daunting, it is clear that the road map being developed by NASA allows for testing from the laboratory bench top to small scale space-to-Earth demonstrations. The time is now to build a technology demonstrator to study the effects of the system on the environment. Failure to address the environmental issues related to the large scale development of space solar power could result in a polarized situation with popular and powerful environmental groups and ecologists on one side and the SSP supporters on the other. This polarization could make the adoption of SSP very difficult, unnecessarily expensive, or impossible.

We must challenge our leaders to make a significant investment in renewable energy sources. Currently, the United States government funds substantial subsidies to the nuclear power and oil industries with minimal support for renewables. If this is going to change, then citizens must be the change agent. Citizens must demand that their elected leaders meet the energy needs of a 21st century planet in an environmentally friendly manner.

Do you think that the concept of limited resources and limited possibilities would change if each of us had the chance to stand on the Moon and look back at the Earth? When you see our beautiful blue Earth suspended in a sea of black — would you start to think out of the box? Space Solar Power lays a foundation far promoting the fact that we live in an open system. Space contains resources and riches that should be harnessed and used to benefit all humankind. While you look back at our planet, you would also notice the impact of humans on the Earth. We can live in harmony with our planet. We can do this now. This technology may also represent the foundation for permanent human settlement in space. As space markets develop and grow, and as the average Earth citizen becomes aware of space as a place, we will see human expansion into our solar system. Today, Space Solar Power is only an idea on paper — the time has come to transform this idea into reality.

# Chapter 15
# *Space Solar Power Systems For The 21st Century*

by
Peter E. Glaser, Ph.D.
Lexington, MA, USA

In 1981 Buckminster Fuller posed one of the most challenging questions in history:

"How do we make the world work for 100% of humanity in the shortest possible time through spontaneous cooperation, without ecological damage or disadvantage to anyone?" [1]

One of the most promising answers to this question is the harnessing of the energy of the Sun in space with Space Solar Power Systems (SSPS).[2] This inexhaustible global energy supply option could begin to meet the escalating demands of the 6 billion people already living on Earth, and reduce drastically the consumption of fossil fuels and their acknowledged threats to the ecology.

The applicable principles underlying SSPS are based on scientific demonstrations in the 19th century, and the successful accomplishments of increasingly complex and demanding engineering capabilities for space missions including lunar exploration and the construction of the International Space Station in the 20th century.

The concept of SSPS has been validated by studies undertaken by the international technical community, and supported by academic institutions, industry and governments. The results of these studies are reported in the substantial literature on the associated technical, economic, ecological and societal issues.[3]

There is a growing consensus that SSPS could deliver sufficient energy in the form of electricity for most conceivable future human needs thereby:

- Increasing the standard of living of all inhabitants on Earth,
- Stabilizing population growth,
- Safeguarding the ecology of the Earth,
- Averting potential global instabilities caused by efforts to control increasingly scarcer terrestrial energy resources, and
- Enabling the development of a space-faring civilization.

The global availability of fossil fuels is finite, with world oil production projected to peak before the first quarter of the 21st century.[4] Global oil production beyond mid-century is projected to drop significantly. In the words of Aldous Huxley:

"Facts do not cease to exist because they are ignored."

So far nuclear power has not achieved its hoped for potential for reasons that are widely acknowledged, and controlled nuclear fusion on Earth remains a challenge to physicists with no agreed upon endpoint in sight.

The nearly inexhaustible fusion process occurring in the Sun can be tapped to meet all foreseeable energy requirements on Earth by converting solar energy directly into electricity in Earth orbits, on the Moon, or other suitable locations in space, and beaming the energy to Earth in preferred wavelength regions. Receiving antennas constructed on land or on the ocean,[5] where energy beams can be safely and very efficiently converted 24 hours a day, would provide electricity to users globally.

The International Space Station makes extensive use of solar power.
The photocell veins occupy a greater area than the station itself.

Feasibility studies, as well as technical, economic, environmental, and societal assessments, and laboratory and field demonstration projects by universities, research institutes, industry, and government agencies, in China, Europe, India, Japan, Russia, Ukraine and USA, have been performed over the span of three decades.

The first international meeting on SSPS was held in the Netherlands in 1970.[6] Frequent meetings were held during the NASA and US Department of Energy: *Satellite Power System Concept Development and Evaluation Program*,[7] at International Astronautical Federation Congresses, and in conjunction with government agencies, international professional society meetings, and at United Nations conferences. As a result there exists an extensive, and globally available literature on the SSPS power supply option.

Several countries have been engaged in wide ranging studies of SSPS, including China, European Union, Japan, Russia, Ukraine, and USA. For example, NASA[8] has stated:

"...The vision for the exploration and development of space is one of a multi-decade expansion of knowledge and the space frontier, which enabled the development of new capabilities and new infrastructure. Multi-megawatt SSPS technologies and systems would make possible breakthrough improvements in projected future space missions. Moreover, in the long term, these technologies might enable the commercial application of SSPS that could deliver energy to terrestrial markets at competitive prices."

The results of 30 years of wide ranging studies by academic institutions, industry groups, professional societies and government agencies are leading to a growing consensus that SSP will:

○ be compatible with the ecology of the Earth
○ meet applicable international safety standards
○ serve as an economically justifiable global energy supply system
○ be in compliance with the international legal and regulatory framework for space operations
○ be acceptable to society

In a broader sense, the SSPS represents a major and meaningful step towards extending peaceful human activities beyond the confines of the Earth's surface, while at the same time contributing to the inevitable transition to an inexhaustible source of energy.

The worldwide interest in and widespread activities to advance SSPS development, demonstrate that it is feasible to place increasing reliance on obtaining energy from the Sun for use on Earth. There is every reason to expect that the benefits of this inexhaustible energy source will be made globally available by the middle of the century.

Intermediate steps of commercial interest leading to SSPS include:

- Long endurance aircraft and dirigibles powered by wireless power transmission from antennas on the ground (demonstrated in Canada and Japan)[9,10]
- Power relay satellites to access remote terrestrial energy sources, and to beam power to receiving antennas across intercontinental distances (USA).[11]
- Deployment of a large solar energy reflector in orbit by Russia, 1993

SSPS should be available on a global scale, and increase the opportunities for all nations to benefit from the inexhaustible energy available in space in accordance with the UN Treaty on: *Principles governing the Activities of States in the Exploration and Use of Outer Space, including the Moon and other Celestial Bodies.*

This Treaty implies, that the SSPS, which must be operated only for peaceful purposes, enjoys the same legal protection as is accorded to installations of power generation systems on Earth which are carried out for the benefit of and in the interest of one or more nations. In principle the SSPS is no different than a terrestrial power plant, a refinery, or a tanker at sea.

The world Administration Radio Conference, responsible for assigning frequencies to all types of space operations has been made aware of the preferred frequencies for SSPS, i.e.: Allocation of the optimum frequency for wireless power transmission from space to Earth in the Scientific and Medical bands.

The opportunities represented by SSPS and its societal implications are profound. It is a unique approach for resolving the major unresolved issues concerning: 1) the future global supply of energy, 2) compatibility with the ecology, 3) equitable distribution of energy, and 4) control of the energy supply by a country or a coalition of countries.

The challenge is to engage in an international program of concerted actions by a broad coalition of potential stakeholders in the successful development and implementation of SSPS. These include a wide range of industries,

This early concept for one element of a solar power station shows a solar heater for helium, electric generators and waste heat radiators.

government agencies, international organizations, including the United Nations, and private and public interest groups who can collaborate to achieve the benefits of power from space for all inhabitants of Earth.

To derive the maximum benefits from a global SSPS, policies will have to be adopted which will be acceptable to United Nation member states, and lead to the formation of the most appropriate international organization for SSPS development and implementation in compliance with the legal and regulatory framework for space operations.[12]

In 1981, Gerard K. O'Neill cautioned us:

"...The knowledge that is required about the physical world, both in the natural and life sciences, is unique in that it builds on all that has gone before and so grows continually. That gives humanity an ever increasing power, and we therefore feel, quite properly, a responsibility to use it wisely. All too easily many of us feel that responsibility in negative terms only, and an obligation to dig in our heels and resist all further change. But the opportunities that now lie before us to shape this next century better than the last depend to a great extent on science and its applications through engineering, and we fail in our duty if we do not courageously seize these opportunities to improve the human condition."[13]

SSPS provides a unique opportunity to access the unlimited energy of the Sun for use on Earth to overcome the unresolved energy supply issues associated with current energy sources, and to ensure the continued progress of the global community in the 21st century and beyond. Now is the time to press forward with adequate resources, and carefully crafted plans to ensure that the Earth will remain a hospitable planet for all forms of life, free humanity from dependence on fire, and allow looking to the future with renewed confidence.

# Chapter 16
# Asteroid Resources, Exploitation, And Property And Mineral Rights
## — or —
# Keep Your Laws Off My Asteroid

by John S. Lewis
Professor in the Lunar and Planetary Laboratory and
Co-Director of the Space Engineering Research Center
University of Arizona

## Introduction

In recent years a number of suggestions regarding the economic exploitation of the energy and material in space have been offered by authors as diverse as Konstantin Tsiolkowsky, Robert Goddard, Arthur C. Clarke, Herman Kahn, Gerard K. O'Neill, Lyndon B. Johnson and Julian Simon. Among proposed resources to be exploited the most commonly cited are:

1) solar energy, from space based photovoltaic cells and microwave transmission of power to antennas on Earth (Solar Power Satellites), to provide clean, cheap energy in vast amounts,
2) the isotope helium-3, a constituent of the atmospheres of Uranus and the other giant planets, and a very minor constituent of lunar regolith, for use as clean fusion fuel on Earth,
3) oxygen extracted from lunar or asteroidal rocks and minerals for use in life support systems and as the oxidizer for rocket engines departing from the Moon, and
4) water from near-Earth asteroids, Martian permafrost, or lunar polar ice deposits for use in life support and rocket propellants.

One important benefit of the use of space derived energy and resources is the enormous positive environmental impact of off-loading mining and energy production from Earth's surface. Another is the prospect that large scale future activities in near-Earth space may be made far less expensive by the use of materials that are already available in space, that therefore need not be launched at great expense out of Earth's gravity well. For logistical reasons, and because of their great resource richness and diversity, the near-Earth asteroids (NEA) have emerged since the first edition of *The High Frontier* as the most attractive source of materials for export (return to near-Earth orbits). For various reasons — most often associated with the energy costs of landing and takeoff, and the lower quality of the resources — profitable export of materials from massive bodies such as the Moon and Mars is far less feasible than is local use.

The idea that very large amounts of money might be made from these space resources has stimulated interest in legal, political and diplomatic circles. All too frequently the motivation seems to be to find a way to extract money from these innovative ideas without actually doing anything to further them. This chapter attempts to present the issues in a more positive light. It briefly surveys the significance, threat and promise of the near-Earth asteroids, with emphasis on the legal and regulatory regime surrounding their economic utilization. Current treaty obligations are reviewed and some suggestions are made regarding legislation and treaty language which would encourage this enormous new arena of economic activity.

## The Promise and the Threat

Recent developments in the astronomical search for near-Earth asteroids, and in geological studies of impact features on Earth and other planets, have presented us with a disturbing vision of the threat of disaster visited upon Earth by the impact of a comet or asteroid. This information shows that, if a large asteroid, one massive enough to threaten human civilization, is on a course to collide with Earth 100 years from now, then that body would almost certainly be one that we have not yet discovered. It also shows that virtually all such large bodies can easily be discovered within the next few decades by a systematic, globally coordinated search and characterization program that would cost less than a single small space mission. With such a search program to give us adequate warning of a threatened asteroid impact, we would then have ample time to design, build, test and deploy an effective defense against the threat.

That same search and characterization program provides us with an exceptional opportunity. Many of the most dangerous asteroids have orbits that are remarkably accessible from Earth: they cross or graze Earth's orbit about the Sun, and can be reached, orbited, and landed upon more readily than any other bodies in the Solar System. Fully a quarter of the near-Earth asteroids, of all classes, are easier to land on than the Moon — a given booster rocket could soft land a larger payload on any of these than it could on the Moon. Further, these bodies have such feeble gravity that departure from them to return to Earth is vastly easier than departure from the Moon. Return of samples from the near-Earth asteroid Nereus, for example, requires a departure speed as low as 60 meters per second (135 mph), whereas departure from the Moon to return to Earth requires a speed of about 3,000 meters per second. The amount of energy (or propellant) required per ton of returned material is 2,500 times as large for lunar missions as it is for Nereus missions. For this reason, export of bulk materials from the Moon to Earth could make sense only for fabulously valuable materials.

Because of the relative ease with which asteroid derived materials can be returned to Earth orbit, very large masses of materials may be moved. Logistical studies suggest that each ton of equipment launched from Earth to a near-Earth asteroid can return 100 tons of material to Earth orbit over the operational lifetime of the vehicle. Thus, assuming a launch cost of under $500 per pound from the surface of Earth, and using this 100:1 leverage factor, materials such as propellants or metals could be made available in Earth orbit for a few dollars a pound. The materials needed for future space transportation and construction would then be comparable in expense to those used in high quality residential construction here on Earth.

In addition to these very favorable energy and logistics considerations, asteroids are attractive targets for a wholly different reason: they're rich in "cheap" materials, such as water and steel, that are of great value and utility in space, but outrageously expensive to launch from Earth. Further, the large majority of near-Earth asteroids contain high concentrations of extremely valuable precious and strategic metals, such as platinum, osmium, iridium, rhenium and palladium, and semiconductor components such as germanium, gallium, arsenic, antimony, tellurium and indium. The Earth surface market value of this fraction is roughly $10,000 per pound, and the concentration, of platinum for example, is higher in the average NEA than in the best known terrestrial ore deposits. Economic studies of retrieval of these materials suggest that they might compete favorably with terrestrially derived materials. Experiments now underway in the laboratory of Prof. Henry Freiser at the University of Arizona — funded by FINDS, a visionary private foundation with interest in the commercial development of space — are seeking simple, effective means to extract and separate many of these valuable resources from meteorites that are authentic samples of asteroids.

The idea of extracting asteroidal materials for commercial use is not a new one — the great pre-Revolutionary Russian rocket visionary Konstantin Tsiolkowsky proposed the exploitation of Belt asteroid resources as early as 1903. The father of practical rocketry, the American physicist Robert Goddard, in 1918 wrote in an essay entitled *The Ultimate Migration* describing interstellar ships made from asteroids conveying our remote descendants away from the death throes of the Sun. The idea of mining asteroids in the Belt was so visionary that Goddard sealed his manuscript away in an envelope labeled "Special formulas for silvering mirrors", where it languished unread for over 60 years. Goddard's reticence is understandable; the technologies required to carry out such ambitious schemes did not then exist, whereas critics and mockers, incapable of understanding his ideas, were legion. Later, in O'Neill's time, knowledge of the near-Earth asteroids was in such a primitive state that he could see only one accessible source of space resources: the lunar surface. But today, with the astronomical knowledge and technology of the year 2000, the dream of mining resources in space can be made a practical reality.

The keys to successful import of materials from space are:

1) lower launch costs
2) careful choice of exploitation targets to favor those that are most accessible and have the richest resource concentrations, and
3) minimizing the complexity of the operations to be undertaken by mining and extraction vehicles so that artificial intelligence, not human presence, can be employed.

The question of how to lower the cost of access to space from about $5,000 per pound to a few hundred dollars per pound is also an old one. Exactly a century ago, Tsiolkovskii wrote a science fiction novel in which the first successful manned venture into space was carried out in the year 2000 . . . by a consortium of industrialists, scientists and technologists funded by what can only be described as venture capital. Today we see nearly two dozen companies, funded by venture capital and equipped with exciting ideas and the most modern aerospace and electronic technology, competing to lower the barriers in the way of the massive development of space.

The inspiration for current mission concepts comes from Alan Binder's *Lunar Prospector*, originally planned as a private venture but eventually partially captured by NASA. The Space Development Corporation's much discussed *Near-Earth Asteroid Prospector* (NEAP) mission to the asteroid Nereus is merely the first example of a deep space mission that embodies the principle of privatizing space activities. The role of NEAP is primarily to provide a science platform for flying NASA funded instruments at low cost, and to sell data gathered by privately funded instruments to NASA. It can in fact, however, be regarded as an authentic prospector — a searcher for useful resources. A second step would be to land processing experiments on the surface of an asteroid and demonstrate small scale production of water or metals. Since return of materials from so many NEAs is exceptionally easy, both virgin surface samples and processed materials may be automatically returned to Earth more easily than samples could be returned from the Moon (a technique demonstrated successfully by the Soviet Union on the unmanned Luna 16, 20 and 24 missions in the 1970's).

Given the accessibility of NEAs and the diversity of their resources, it's useful to give some idea of the amounts of resources available. Perhaps the first example of note is the metallic near-Earth asteroid called Amun. This asteroid is about 2,000 meters in diameter. If it were to strike Earth, it would deliver a devastating blow of 10 million megatons (10 teratons) of TNT, several thousand times the explosive power of a nuclear world war. Amun, the smallest metallic asteroid of the several dozen known, contains nearly 10 times as much metal as the entire amount of metals mined and processed over the history of mankind. A conservative estimate of the market value of this asteroid is 40 trillion dollars.

The entire NEA population, which is very diverse in its chemical and physical properties, contains vastly more material than this. An estimate of the overall composition of the NEA population is shown in Table I. It's not difficult to estimate how much of each of a wide variety of commodities, such as water, carbon, nitrogen, metals, phosphorus, etc., is required in circulation to maintain one human adult. From these figures, we may estimate how many people could be supported indefinitely by the resource wealth of the NEA population in a fully recycling regime, powered by the Sun. According to Table I, the number is probably close to 14 billion people. Nitrogen, principally in its role as a fire suppressing dilutant of atmospheric oxygen, appears to be the limiting resource. One of the most remarkable aspects of Table I is that the proportions of materials needed by civilized human beings are closely similar to the proportions found in asteroids.

## Table I: Resources of the Near-Earth Asteroids[*]

| Commodity | Mass present among NEAs ($10^{15}$ g) | Per Capita Inventory (g/person) | Population Sustainable by NEA Resources (billion people) |
|---|---|---|---|
| Silicates | 2,500 | 140,000,000 | 17.8 |
| Ferrous metals | 300 | 20,000,000 | 30.0 |
| Fe in oxides | 300 | — | — |
| Cement | 60 | 10,000,000 | 6.0[1] |
| Industrial CaO | — | 2,000,000 | 30.0 |
| Phosphates | 10 | 2,000,000 | 5.0[2] |
| Water | 300 | 10,000,000 | 30.0 |
| Carbon | 100 | 1,000,000 | 100.0[3] |
| Nitrogen | 10 | 700,000 | 14.0[4] |
| Sulfur | 60 | 1,200,000 | 50.0 |
| Sulfides | 150 | 1,200,000 | 125.0 |

[*]  The near-Earth asteroid population is a renewable resource that replaces itself about every 30 million years. Note that the only true "consumable" in this fully recycling system is solar power.

[1]  Cement is of dubious utility in space, except for wholly internal (non-vacuum tight) construction.

[2]  Phosphate fertilizer usage on Earth is predicated upon toleration of massive loss in runoff from fields. In a 100% recycling regime, the required inventory could easily be 10 times smaller.

[3]  Carbon inventories assume 1,000 g of plant carbon per gram of human carbon.

[4]  Nitrogen inventories assume 1,000 m$^3$ of habitat volume per person.

But the NEAs are only a small part of the picture. Most NEAs follow orbits that take them out to the heart of the asteroid Belt, between the orbits of Mars and Jupiter, at aphelion. Thus a processing unit landed on a typical NEA will get a free round trip to the Belt and back on each trip around the Sun (typically, once every three to five years). The asteroids that reside in the Belt make up a very large population, with a mass that is roughly a million times as large as the total mass of the NEA population at any one time. Table II summarizes the total amount of resources available in the Belt in a manner similar to the NEAs in Table I. The conclusions are staggering: the materials available in the Asteroid Belt would suffice to maintain indefinitely a human population of at least 10,000,000 billion people — about one million times the maximum carrying capacity of Earth. Assertions that we are running out of resources reckon without 20th and 21st century technology. The supply of resources available to a spacefaring humanity is effectively infinite. But there is no possibility of either returning all that material to Earth (there's enough steel in the Asteroid Belt to build a steel frame building 8,000 stories tall covering the entire land area of Earth) or of accommodating so many people on one planet.

In the likely order of development, water is the first asteroidal resource worthy of attention. Its uses as a propellant and as life support materials are obvious. Second in order would be native ferrous metal alloys, of which the major components — iron and nickel (and possibly cobalt) — would be retained in space for use in the construction of space based facilities such as Solar Power Satellites. The rare and very valuable precious metals and semiconductors in asteroidal metal alloys are worth returning to Earth. Dr. Jeffrey S. Kargel has explored the effects of large scale import of these materials from space on the market size and prices of these commodities on Earth. He concludes that prices will decline less rapidly than the rate of supply increase because new uses will be stimulated by the price decreases.

## Table II: Resources of the Asteroid Belt[*]

| Commodity | Mass present in the Belt ($10^{21}$ g) | Per Capita Inventory (g/person) | Population Sustainable by Belt Resources (billion people) |
|---|---|---|---|
| Silicates | 2,500 | 140,000,000 | 17,800,000 |
| Ferrous metals | 300 | 20,000,000 | 30,000,000 |
| Fe in oxides | 300 | — | — |
| Cement | 60 | 10,000,000 | 6,000,000[1] |
| Industrial CaO | — | 2,000,000 | 30,000,000 |
| Phosphates | 10 | 2,000,000 | 5,000,000[2] |
| Water | 300 | 10,000,000 | 30,000,000 |
| Carbon | 100 | 1,000,000 | 100,000,000[3] |
| Nitrogen | 10 | 700,000 | 14,000,000[4] |
| Sulfur | 60 | 1,200,000 | 50,000,000 |
| Sulfides | 150 | 1,200,000 | 125,000,000 |

[*]  Assumes full recycling and full reliance on solar power.
[1]  Cement is of dubious utility in space, except for wholly internal (non-vacuum tight) construction.
[2]  Phosphate fertilizer usage on Earth is predicated upon toleration of massive loss in runoff from fields. In a 100% recycling regime, the required inventory could easily be 10 times smaller.
[3]  Carbon inventories assume 1,000 g of plant carbon per gram of human carbon.
[4]  Nitrogen inventories assume 1,000 m$^3$ of habitat volume per person.

The potential customers for the materials that remain in space would include government agencies requiring propellants for injecting equipment into geosynchronous orbit, for orbital station keeping, or for departure out of Earth's gravity well, such as both unmanned and manned Mars missions. In addition, civil traffic bound for geosynchronous orbit would be an important market for propellants. Metals would be of use for construction and shielding of large structures, of which the most obvious commercial example would be Solar Power Satellites. Since, according to a recent NASA study, SPS's are close to being economically viable even with Earth launch ("uphill" transport) of all of their components, a scheme that derives their most massive, low tech parts from non-terrestrial metals (brought "downhill" to Earth orbit) holds promise of making them highly competitive economically.

Some words are also in order regarding the environmental impact of space resource use. In general, any industrial activity that can be offloaded from Earth eases the environmental burden on Earth's biosphere. Enormous proportions of Earth's environmental troubles are related to the mining, refining, transportation and use of fossil fuels. National policies aimed toward reliance on Solar Power Satellites would mitigate all these problems, simultaneously making the United States the world's largest exporter of energy and energy technology, and extending the usefulness of our limited petroleum reserves far into the future, dedicating crude oil to petrochemical production, not combustion. Finally, such an approach reduces or eliminates American dependence upon foreign sources of crude oil.

But this entire scenario is dependent upon a stable and rational legal and regulatory system in which investors will have reasonable assurance that the fruits of their ingenuity and investment will not be arbitrarily obstructed, or even confiscated. Unfortunately, history records that the exploitation of vast mineral wealth on Earth's ocean floor was rendered impossible by the absurdly restrictive *Law of the Sea Treaty* that essentially stripped any successful entrepreneur of half of his discoveries, half of his profits, and all of his proprietary technology. This communal dream appealed to many nations that lacked the scientific knowledge, technical capability, and economic power to enter the business for themselves. It just seemed easier to take from those who were capable than to participate in any positive way. The *Law of the Sea Treaty* had an unintended consequence: rather than assisting its advocates in expropriating the fortunes and property of successful entrepreneurs, it instead killed the goose before it could lay its first golden egg. And this infamous treaty set the tone and pattern for early attempts at writing space law.

## Present Legal Regime

The basis for more recent treaties was laid by the 1967 *Treaty on Principles Governing the Activities of States in the Exploration and Use of Outer Space, Including the Moon and Other Celestial Bodies*, usually referred to as the *Outer Space Treaty*.

Article I states that: "The exploration and use of outer space, including the Moon and other celestial bodies, shall be carried out in the interests of all countries, irrespective of their degree of economic or scientific development, and shall be the province of all mankind." It further states that the exploration and use of celestial bodies shall be done "without discrimination of any kind".

Article II proclaims: "Outer space, including the Moon and other celestial bodies, is not subject to national appropriation by claim of sovereignty, by means of use or occupation, or by any other means."

Article III says that "States Parties to the Treaty" shall carry out their activities in accord with international law.

Article IV, reserving space for peaceful activities, forbids nuclear or other weapons of mass destruction to be installed on any celestial body or "stationed" in space in any manner.

Article V commits the States Parties to assist astronauts in distress, which simply expresses the principle of the 1968 *Agreement on the Rescue of Astronauts, the Return of Astronauts and the Return of Objects Launched into Outer Space*.

Article VI binds each signatory State to take responsibility for "national activities in outer space, including the Moon and other celestial bodies, whether such activities are carried on by governmental agencies or by non-governmental entities ..."

Article VII assigns liability for damage by space missions to the originating State, a reaffirmation of the principle of the 1972 *Convention on International Liability for Damage Caused by Space Objects*.

Article VIII affirms the 1975 *Registration of Objects Launched into Outer Space*, and Article IX affirms the principle of cooperation and mutual assistance in deep space developed in the 1970 *Declaration on Principles of International Law Concerning Friendly Relations and Cooperation Among States in Accordance with the Charter of the United Nations*.

Article X calls for the free opportunity to observe launches and flight of the spacecraft of all nations, which is of course routinely done without any involvement of the observed party, and never done with any meaningful ability to determine the contents and function of the payload.

Article XI reiterates the principle of international (UN) registration of launches "to the greatest extent feasible", without any requirement for prompt or substantive reporting, following the 1975 convention on *Registration of Objects Launched into Outer Space*.

Article XII requires that all "stations, installations, equipment and space vehicles on the Moon and other celestial bodies shall be open to representatives of the other States Parties to the Treaty" and to other States on the basis of reciprocity.

Article XIII places intergovernmental agencies under the same Treaty obligations as individual States. Articles XIV-XVII concern ratification and amendment of, and withdrawal from, the treaty, and asserts equal authenticity of the text in several languages.

In general, the *Outer Space Treaty* assumes that the domination of space activities by governments that characterized the 1960's was a permanent feature, and that commercial ventures could be dismissed without individual consideration by simply assigning them to the nearest available government. Private property is not even mentioned. The 1967 Treaty, high in nobility of intent and short in substance, was ratified by every spacefaring nation, and reflects a very broad international consensus. For our purposes, the most important single feature of the *Outer Space Treaty* is the assertion in Article I that uses of space are "the province of all mankind." The interpretation of this phrase is almost infinitely varied, reflecting the political view of the interpreter. The most basic interpretation is that everyone has the right to participate in exploration and use of space, and no one can be denied that right. The most ambitious would require representation of (or permission from) every nation on every mission. The fact that the treaty is generally acceptable to a wide range of nations with different and often contradictory political ideologies suggests its fundamental lack of objective meaning.

The most relevant document concerning exploitation of asteroids is the 1979 *Agreement Governing the Activities of States on the Moon and Other Celestial Bodies*, more commonly known as the *Moon Treaty*. Draft materials for this treaty and an account of the negotiations leading to it are included in a U.S. Government Printing Office publication, 59-896 O (1980). The final agreement, adopted and opened for ratification by States by General Assembly Resolution A/RES/34/68 bearing the same name, consists of 21 articles.

Article I includes all other celestial bodies with the Moon as being covered by this agreement, excepting meteorites that fall to Earth "by natural means".

Article 2 echoes Article III of the 1967 Treaty.

Article 3 bans military activities and installations (as in Article IV above), Article 4 repeats Article I, Article 5 repeats Article XI on notification, and Article 6 on "freedom and equality" reflects other language from Article I. It also guarantees the right to remove samples for scientific purposes.

Article 7 prohibits "the disruption of the existing balance of the environment" and calls on States to avoid "harmful contamination" of the body as well as "harmfully affecting the environment of Earth".

Article 8 guarantees the right to land, establish bases and travel over the surfaces of these bodies.

Article 9 enjoins these States to "use only that area that is required for the needs of the station".

Article 10 renews the commitments of the 1968 Agreement on the safety of astronauts.

Article 11, by far the most important for our purposes, states that "the Moon and its natural resources are the common heritage of mankind". Further, "the Moon is not subject to national appropriation by any claim of sovereignty, by means of use or occupation, or by any other means". This article echos the 1967 treaty. Neither the surface, subsurface, nor natural resources of the Moon "shall become property of any State, international governmental or non-governmental organization, national organization or non-governmental entity or of any natural person." Installation of equipment or bases does not create any claim of ownership. "The States Parties to this agreement hereby undertake to establish an international regime, including appropriate procedures, to govern the exploitation of the natural resources of the Moon as such exploitation is about to become feasible." States Parties are required to report any natural resources discovered on any celestial body to the Secretary General of the UN. Of course, *anything* in space is arguably a resource — and no resource of any kind has even been reported. In closing, Article 11 proclaims that "The main purpose of the

international regime to be established shall include:
a)  the orderly and safe development of the natural resources of the Moon,
b)  the rational management of those resources,
c)  the expansion of opportunities in the use of those resources, and
d)  an equitable sharing by all States Parties in the benefits derived from those resources, whereby the interests and needs of the developing countries, as well as the efforts of those countries which have contributed either directly or indirectly to the exploration of the Moon, shall be given special consideration."

Article 12 allows ownership of devices and stations placed on the Moon, Article 13 obligates States to report crashes of hardware of any nation on the Moon, and Article 14 places full responsibility for all activities on the Moon upon the State from which that activity originates.

Article 15 requires that every installation on the Moon be open to inspection visits from any other State, and the remaining Articles refer to the role of international governmental organizations and to the ratification and amendment of the agreement.

It is clear that the nations contributing to this document accept absolute centralized control and common ownership of these resources, with required "sharing" (confiscation) of much of the proceeds of the work in exchange for nothing. The treaty represents a particular vision of reality that seems to offer nothing for any state capable of conducting operations on the Moon. The United States and essentially every other nation with spacefaring capabilities are signatories to all of the other treaties and agreements cited above. However, with almost equal unanimity, these same nations have failed to ratify the *Moon Treaty*. Many other nations, with nothing to lose, have hastened to ratify it, hoping perhaps for some windfall of risk free income from the labor, investment and invention of others. They want interest but have no principal; they desire dividends without the trouble of buying stock. They would be far better advised to ask how they can help out with the endeavor and earn a share in its profits. Austria, which ratified the *Moon Treaty* at a time when it had no space program, is now a member of ESA. It is amusing to speculate whether Austria would again ratify the Treaty if the issue were raised today.

## Suggested Legislative Actions in Support of Space Development
Perhaps it would be more fruitful to ask what legal, regulatory and economic regime we *want* to have, rather than what others have wanted in the past. Which matters should be subject to international regulation? What functions should be carried out by national governments, and which should be left to the free will and choice of the private and public spacefaring parties themselves? I would like to offer a few thoughts on the division between governmental and commercial roles, especially to define and suggest the appropriate role of government. As I see it, opening the space frontier, like the opening of the American West, can be assisted or hindered by governmental actions. Just as with the establishment of railroads in the West, government assistance, not domination, is useful in several areas:

1.  The government can support private space endeavors by buying scientific data, in effect *privatizing many research missions*. The cost to the taxpayer is negative: it saves money. The NEAP mission is but the first example of this approach, offering a cost structure that reflects competitive commercial practice rather than the inefficiency that is characteristic of monopolies. NEAP and its likely near future competitors will offer prices about a factor of four lower than the cost of doing a comparable mission from within the government. This improvement is above and beyond the already substantial cost reductions brought about by Mr. Goldin. This goal will largely be achieved under the provisions of the Commercial Space Act of 1997.

2.  The government should support the development of low cost space transportation systems by *buying launch services competitively from private vendors*. The historical domination of launch services by government agencies has been the chief factor in preventing launch costs

from declining over the past 30 years. If properly conducted, this initiative also would cost the government nothing, and indeed promises very substantial cost savings on ambitious future missions. A good start on this initiative was made by the Launch Services Purchase Act of 1990.

There are several important ways in which governmental agencies can further space development through support of basic technology development:

3.  The government, through NASA, should take a leading role in *developing key electronics technologies* to assist early economic development of space, as it did in the early days of NASA with communication, navigation, Earth resources and weather satellites. The present priority list should include teleoperation and telepresence. Especially useful would be the development of extremely capable, flexible and highly autonomous computer control systems and expert software systems to permit both very distant NASA exploratory missions and future commercial operations to function, as far as feasible, independently of detailed control from Earth. Given the high level of sophistication of the American computer industry, such a development program should not be expensive, and would have obvious spin-offs in hazardous mining, industrial and military environments on Earth.

4.  The government should play a major role in *developing critical propulsion technologies*. At a bare minimum, solar thermal and solar sail propulsion systems should be developed and placed in the public domain. Here the Air Force's history of studies of solar thermal propulsion systems make them an obvious participant. Nuclear thermal rockets, which play a potential major role for missions requiring relatively high thrust levels or operation at great distances from the Sun, are not essential for early exploitation of asteroidal resources. Flight testing of demonstration scale solar thermal rockets and solar sails, and of computer algorithms for control of the pointing of their mirrors, could be carried out for tens of millions of dollars. The Russian orbital test of a steerable solar mirror (*Znamya*) in November 1998 was the first flight experiment relevant to this capability.

5.  The government should take the lead in *developing technologies for extraction and processing of the most important mineral resources*. The technologies that should be tested in microgravity at the Space Station, or elsewhere, include crushing and magnetic separation for extraction of native ferrous metal alloys, and extraction of water from ice lenses, permafrost, hydrated salts and clay minerals. Initially, water may be used directly in solar thermal or nuclear thermal rockets, but it is clear that, in the longer term, conversion of the water into cryogenic propellants will be highly desirable. Electrolysis of water into hydrogen and oxygen, and liquefaction of both gases to make liquid rocket propellants, must also be adapted to operation in very low (or artificial) gravity. Many of these technologies are extensions or adaptations of familiar Earth surface processing technology to high vacuum, microgravity environments. Metals mining and processing concerns have extensive Earth related experience, but require expert assistance from universities and government research centers in selection of their targets, and from the government and the aerospace industry in adapting their experience to space operations. Given the present precarious financial state of the aerospace industry, their participation in the experimental stage of such an endeavor would almost certainly require government support. As recently as 1993, NASA was spending over $2.5 million per year on research into the use of non-terrestrial resources. This endeavor has been heartily endorsed by the NASA Administrator in many public addresses, but nonetheless funding has essentially disappeared. This is but one example of an area in which NASA's leadership is far more visionary than its execution. NASA funding for asteroid related resource work was terminated several years ago, and the effort has only been restarted recently on a small scale, supported by a visionary private foundation, FINDS, associated with the Space Frontier Foundation. A governmental (NASA, U.S.G.S., Bureau of Mines) level of funding of a few million dollars per year would permit active pursuit of a number of extraction, processing, and fabrication techniques, including flight testing on small, low cost boosters.

6. The government should take the lead in *purchasing products produced in space for use in space*. The leading example is rocket propellants. The costs of ambitious deep space missions are raised enormously by the cost of lifting the required fuel load for the outbound trip out of Earth's gravity well — at a cost of several thousand dollars per pint. Instead, all outbound missions passing through Low Earth Orbit could be refueled with high performance asteroid derived liquid hydrogen and liquid oxygen from a "gas station" in Low Earth Orbit, perhaps an adjunct to the Space Station. Since the government would simply purchase high performance, cheap space derived propellants that are competitive with Earth launched propellants, this activity also saves the taxpayer money. Up-front investment in experimenting with the transfer of cryogenic propellants in microgravity would be required before any benefits could be realized. It is interesting that the Soviet Union has used on-orbit propellant transfer routinely and without incident since the 1970's, whereas the United States has never developed the capability. Russian collaboration could be very useful in this area. Beyond propellant manufacture, the most promising single space resource for immediate exploitation is native ferrous metals, which could be used for large scale space construction. Using asteroidal metals to build Solar Power Satellites appears to be the easiest route to energy self-sufficiency for the United States, Japan and Europe.

7. The United States has a unique opportunity, by virtue of its economic strength, scientific superiority and political predominance, for *exercising positive leadership in space*. My approach is that of an unabashed capitalist who believes that the healthiest economy, and therefore the best kind for all citizens, is the one in which free competition allows bright ideas with economic potential to emerge rapidly. We should, if we believe these principles, be outspoken in our support of scientific and technological visionaries even if they happen to be residents of other countries. Many observatories and observers throughout the world are now idle as the result of diversion of funding into new, large telescopes dedicated to deep sky observations. These orphaned facilities could be added to the search effort for modest a cost. In addition, small NASA grants to asteroid search teams in the Southern Hemisphere or Asia would not only assist in hastening the day when we know what is coming, but might also embarrass local governments into exercising their own responsibility to the future safety and economic well-being of their peoples. They have much to gain in the longer run, but will do so only if they see ways to be major participants in the future.

8. Finally, in light of the economic potential of the points discussed above, governments must be prepared to deal with *certifying private claims of ownership and mineral claims on bodies in space*. Governments should establish a formal means of registry of these claims which will assure entrepreneurs that the integrity of their claims will be recognized, a necessary precondition to the raising of large amounts of venture capital. Governments must resolutely avoid international legal entanglements such as the original *Law of the Sea* treaty, with its confiscatory attitude toward profit and new technologies. Such treaties prevent the development of resources that would ease the lot of all mankind. Failure to act may result in damaging actions by international agencies dominated by nations with no present capabilities to operate in space, whose desire is simply to skim off the profits from, or simply punish, those nations more advanced and capable than themselves. By so doing, these developing nations would deny themselves the ability to move into space as launch costs decline over the next decade. They must be made to see that opening up new resources and reducing costs is to their direct benefit, that they too may be direct participants in such activities. As I pointed out earlier, the resources available in nearby space are so large, and the cost of access to space will soon be so low, that the idea of domination by any monopoly is absurd. The most important effect governments can have on space development is to lay a firm foundation for commercial, competitive, private development of space resources.

## Suggested International Regime

A host of important questions regarding international space development remain unanswered:

1) Can individuals and corporations make mining claims in space?
2) If so, with whom?
3) What conditions must be met in order to file a claim?
4) In disputes regarding a mineral claim, in what venue will the dispute be adjudicated?
5) In the event that a claimed mine site is jumped, or an adjudicated agreement is violated, what form of enforcement is appropriate, and by whom is it exercised?
6) What agency, if any, exercises rights of regulation, inspection, and oversight? Who pays for the trip?
7) What precautions can be taken to prevent inspectors from appropriating proprietary technologies?
8) What courts have jurisdiction over disputes between two spacefaring nations that are not signatories of the *Moon Treaty* (none of them are)?
9) Is direct private or corporate ownership possible?
10) If so, who certifies ownership?
11) What degree of presence is required before a claim (either of mineral rights or ownership) can be made?
12) Who enforces property rights?
13) What rights, if any, do the astronomers who discovered, tracked, named and spectrally characterized these bodies have?

For the purposes of the following discussion, we will of course assume that some profitable form of space resource enterprise exists or demonstrably could exist. If this were not true, none of the above questions would be of any interest. "Profitable" means that the proposed space activity can either provide a commodity in some economically meaningful location for lower cost than its competition, or provide much larger amounts of that commodity at comparable prices, or provide a new commodity not already on the market. In all of these cases, it is to the economic advantage of the buyer, the entrepreneur, and the governmental entities that tax profits that such an activity be allowed to flourish. This is true even without allowing for the environmental advantages of relegating energy and mineral mining to space, which positively impacts every person on Earth. It is perfectly true that certain existing interests, especially those that have monopolies on rare materials, would be harmed by such competition based on new resource bodies and new technologies. But this has always been true, and always will be. Successful mining concerns are those that have taken the lead in adopting new technologies and developing new ore bodies. Metals concerns such as INCO will be faced with the choice of trying to compete with space derived resources, or of taking a leadership role in their development. All consumers will benefit — only those corporations ready and willing to adapt will benefit from this new trillion dollar market.

We therefore require that some means be instituted to make space mining possible, on the grounds that it is advantageous to humanity. This in turn requires that the entities who carry out mining, extraction and fabrication of space resources be given a regulatory regime in which investments would be rational. There are enough economic and physical risks associated with space mining so that adding the risk of an unstable or politicized regulatory environment would be fatal. It is therefore essential that mining claims be registered, recognized and enforceable.

The simplest, and in many ways most attractive, scheme would have the United Nations serve as a registry for mineral claims, much in the same way that it maintains a registry of launchings of space vehicles without exercising any control or authority over the activity. The World Court could also provide a venue for claim registry. Another method, inspired by patent law, would allow private individuals or corporations of any nation to register mining claims with their own government, with full mutual recognition of claims. This sounds cumbersome, but in practical reality nations will enter this arena one or two at a time, executing reciprocal agreements with the other relevant governments as the need arises. Adjudication of conflicts would fall within the venue of the World Court. Enforcement would of course be carried out on Earth — the science fiction device of asteroid miners declaring autonomy is not a near term option, as long as space activities are highly dependent on Earth for both equipment and markets. The missing element, of course, is that the World Court lacks any enforcement capability.

There are several levels of presence that might serve as the threshold for making a valid mining claim. They are:

1) at the lowest level, discovery of an asteroid,
2) remote spectral characterization of an asteroid demonstrating the presence of an economically attractive resource (i.e., an ore),
3) physical presence of an unmanned vehicle to document the presence and setting of an ore,
4) physical presence of an unmanned vehicle, documented by sample return to Earth,
5) physical presence of a human crew which proclaims mineral rights or ownership, and
6) presence of an established human settlement.

It is noteworthy that of the roughly 8,000 asteroids discovered and catalogued to date, and of the nearly 1,000 that have been subjected to photometric or spectroscopic study, not one single astronomer has ever registered a public claim of mineral rights or of ownership. This is *prima facie* evidence for a universal consensus that such data do not constitute valid grounds for a claim. Maritime law has provided a precedent for claiming "mining" rights to shipwrecks through reconnaissance by unmanned vehicles that visit and document the "ore". There has been recent commercial interest in the auction of one of the Soviet *Lunokhod* lunar rovers (still resident and inert on the Moon) as the possible basis for a claim of ownership of the terrain surveyed by it in its active career. Some space law theorists have suggested that a higher level of claim be awarded to those who return a sample to Earth. However, there seems to be no clear legal precedent for this requirement, and a host of scientific and engineering reasons to suppose that physico-chemical characterization by spacecraft instrumentation provides all the essential information that could be provided by a returned sample. The need to assay for the abundance of a particular ore, an economic necessity in mineral claim assessment on Earth, is largely irrelevant on homogeneous asteroids. However, as pointed out to me by Tim Pleasant, the terrestrial standards of proving exploitability of a claim and demonstrating an on-going expenditure of funds on exploitation of the ore body may translate directly into standards for recognition of claims in space.

For this reason and others, there seems to be no need to require human presence, since the evidence needed to establish the presence of an ore can readily be acquired without human presence. Further, requiring human presence places an enormous economic barrier in the way of asteroid resource exploitation, effectively preventing private corporations and small nations from participating. Only a handful of superpowers could afford to carry out such a mission, artificially turning space resource exploitation into a monopoly or elitist activity. This is precisely the opposite of the desirable regime, in which many nations and companies can participate, and in which free competition places constant pressure on prices, to the benefit of consumers. Note that manned presence on another celestial body, the Moon, has not resulted in any claim, since the entity that carried out the Apollo project was NASA, a government agency that, by recognized international law, cannot make a claim of national sovereignty. Alan Wasser's suggestion that the United States Congress should pass a law authorizing an "extraterrestrial land claim made by any private entity that has established a true space settlement" is completely inadequate. Imagine if you were not allowed to seek ownership of the land your house was built on until after you (and a number of others) had built your houses and moved into them! In a commercial regime, investors need reasonable prior assurance that their capital is not being poured down a rat hole. Claim recognition absolutely must precede development.

In general, the vast amounts of resources available among the near-Earth asteroids and in the Asteroid Belt suggest that competition for mining claims should not be a common problem. Nonetheless, for any particular resource, such as water, there may be a single asteroid that is significantly more accessible than the others. There would then be a strong incentive to get there first and file a mineral claim that included the entire asteroid. For a 100 meter or 1,000 meter body, comparable in size to an open pit mine on Earth, it would be perfectly reasonable for the first arrival to claim the entire body. On the Moon, filing a mineral claim for, say, the entire Mare Imbrium, would be an absurdity, comparable to claiming all of Kansas or Switzerland as a mine site. Further, on a very small asteroid, minor environmental disturbances caused by one mine (dust, for example) might materially hinder other activities on the same body. Although it's true that the asteroid environment is rapidly self-cleaning (through re-accretion of dust and Poynting-Robertson removal of escaped dust), the short term local effects on visibility may be serious. A reasonable criterion might be to allow claims of all the material of that body within one kilometer of the designated mine site in all directions, excepting any material already claimed in conformance with this criterion. The documentation of the claim must therefore include not only physico-chemical data on the surface, but also a good three dimensional map of the body with the mine site specified. An actual landing at the mine site (not necessarily permanent occupancy) may be mandated, although I don't see any convincing reason to require it. The concept of an ore body has quite different meaning on an asteroid than in usual terrestrial experience. Except for composite asteroids assembled by chance low velocity collisions, most NEAs should be compositionally uniform. The physical state (dust, sand, cobble, boulder, country rock) of the resource is usually a more important determinant of mine site location than the composition, which will usually be very uniform. There will rarely be veins of ore to follow. Further, the dross from extraction of any resource (such as water from a carbonaceous asteroid) will itself be a very valuable resource, containing an abundance of other volatiles and both ferrous and precious metals.

This brings us to the issue of actual private or corporate ownership. A valid claim confers essentially all the benefits of ownership except, perhaps, the ability to sell or license the claim. I would regard a claim that can be sold as the moral equivalent of property. Allowing claims to be sold permits optimization of expertise relating to prospecting and site study, and will help create mine development specialists. I for one would regard mineral claims that carry the right of sale or licensing to be completely satisfactory.

It must be emphasized that no entity that has met the basic requirements of either human presence (NASA astronauts on the Moon) or *in situ* representation by an unmanned spacecraft (American *Viking* and *Pathfinder* and Soviet *Mars* landers on the Martian surface) has ever claimed property rights or national sovereignty. Certainly those individuals who presently purport to be selling tracts of land on the Moon and Mars have no basis whatsoever for claiming ownership in the first place. They lack both the right to sell what is not theirs, and the legal or juridical authority to act as registrars of ownership. The existence of such schemes serves to discredit legitimate mineral rights or ownership claims based on reasonable criteria of presence. Similarly, the existence of *ad hoc* registries for claims, such as that of Prof. Lawrence D. Roberts of the Archimedes Institute, should be taken no more seriously than, for example, the proceedings of a moot court. One can only hope that the existence of this service stimulates awareness of the problem, so that some official mechanism for claim registry may be brought into existence by treaty. The issue of claim registration is made timely and urgent by the impending *Near-Earth Asteroid Prospector* (NEAP) mission of the Space Development Corporation. The time for initiation of a recognized claim registry system has arrived.

# Chapter 17
# *A Conspiracy Of Dreamers*
by
Rick N. Tumlinson

By choosing a non-specialist couple as his letter writers for *The High Frontier* (Jenny is a turbine blade inspector and Edward works in precision assembly), Dr. O'Neill took aim at the heart of the "right stuff" Roger Ramjet space program, and threw open the doors of the frontier to the rest of us. What more human a voice to give his version of our future in space than that of a husband and wife writing home to their family on Earth? What more gentle assault could one mount on the gray metal, military-industrial space program than the story of everyday people living in a small town that just happens to be floating in space?

Dr. O'Neill was not the first to speak of human communities in space. However, he was the first to put all the pieces together and come up with an economically driven scenario that sounded plausible. He was the first to make it understandable to anyone, and most importantly, he was the first to invite those reading his words to go and make it happen.

From Tsiolkovsky to Dandridge Cole, others had posited the idea of cities in space, and others such as Peter Glaser had presented the concepts of space solar power. NASA had myriad concepts for human habitation in space, but they were all designed to provide housing for a few government employees on exploration junkets. After all, exploring space was something to be done for us, not something that we, the great unwashed, might ever hope to take part.

Since the success of NASA's quest for the Moon, the agency had been lifted to a place of mythological proportions in our culture. The cost associated with space made it seem impossible that anyone without the deep pockets of a Federal agency could play the space game. Also, the need to garner public support for its endeavors and keep the funds flowing led to a public relations campaign that played up the very real dangers of space, and the near god-like heroism of those daring to enter its cold blackness. Ironically, right along with the rest of the world, Americans themselves, the descendants of generations of hands-on do-it-yourself pioneers, also bought the idea that space was out of their reach. It was as if the children of pioneers had ceded their history and heritage to an organization of supermen who would literally go out and conquer space for us while we sat comfortably on our couches and praised the state who made it so.

Gerry O'Neill took a national US project, allegedly based on exploration and science, and added the next logical step — Settlement. Yet, he didn't take anything from the proud track record of our national space program. Instead, he built upon its achievements. Then he took it to a new level, by giving the frontier to the people.

He gave us all permission to dream. For the worlds in space that he envisioned were communities just like the ones we inhabit here on the surface. They looked liked our neighborhoods, with green grass, trees and birds. More importantly, they were filled with people just like us.

Unlike many writers, who are content to simply toss their ideas out into the world with no thought of going any further than intellectual masturbation, Dr. O'Neill saw the publishing of *The High Frontier* as only the beginning. He didn't just "talk the talk" as they say, he also "walked the walk."

When he and his wife, Tasha, set up the Space Studies Institute they created a major tool that could be used in the realization of his dream. They may have been doing it to create a new source of funds for space development outside of the government, but by the very nature of its structure and goals they did much more, they also planted the seeds for a conspiracy of dreamers.

It's very clear to me that Dr. O'Neill is the one who started the space frontier movement. From the founding root of SSI sprang dozens of vibrant offshoots such as L5 in the 70's, the Space Frontier Foundation, the ProSpace lobby, the International Space University, the multi-million dollar FINDS endowment and dozens of other groups around the world. His work is acknowledged as a key inspiration for those leading the new space firms, building today's space planes and tomorrow's new space industries.

More subtly at times, the O'Neillian conspiracy has redrawn the image of our nation's future in space. From its branches have fallen the seeds of change in US space policy, and judging from the dramatic changes in focus that we see in our space program and space industries today, those seeds have taken root. Since *The High Frontier* was first published in 1977, many a brave frontier dream has already died, many will still wither, but some will survive, and they will be stronger for the effort, as their roots will be deep. And in the next few years their fruits will lead us to the first harvests of space.

The story of the frontier movement is big and getting bigger. I shall leave it to others to write the exact history of this conspiracy for the stars. Here I will only address a few of the parts of this incredible and still on-going revolution that I personally experienced.

First, I must say this conspiracy predates my entry into the family, and early efforts to spread the word were well underway before my arrival on the scene. In fact, if it involves the leading edge of space frontier technology, SSI and Gerry probably had something to do with it at some time.

One of the best parts of the structure he laid out for the Space Studies Institute was the Senior Associates program. Although his original goal was to create a cadre of dedicated contributors, what in fact happened was the creation of a core group of conspirators for the cause. These were people who were sophisticated about the issues, dedicated, and most importantly, felt partial ownership of the ideas. In a very real sense, they were hands-on transforming those ideas into reality.

The Senior Associates are a unique breed. You see, via SSI, they are actually involved in projects to open up space. They aren't mere fans, nor even activists armed only with policies and ideas. No, SSI's Senior Associates (SA) can legitimately say, "I am involved." Other than amateur radio, this was one of the first times a non-government entity or sanctioned group had participated directly and independently in the opening of space. It made an incredible difference. We were not only helping fund an alternative space program, we were helping pick those projects, and often rolling up our own sleeves to get involved.

The SSI Princeton Conferences on Space Manufacturing provided one of the only venues for new and untried ideas to be rolled out in front of an only partially hostile audience, a function they still hold today. We informally called them the *The Not Ready for Prime Time* conferences. The concept was to leap over the elitist walls of other academic space events and create a renaissance climate open to new ideas. They were the only place I knew of where a medical doctor with an idea for asteroid mining and an engineer from a cement company, who wanted to make Moon bricks, could present their ideas to some of the top names in space and be given a serious review. They were also gatherings of the faithful, a place of sharing ideas and plans to open the frontier.

I recall long nights at SSI conferences, when the SA's would gather and plot out our future on the frontier. From debates over technologies to strategic discussions over the fate of the known universe as we saw it, many from this group became and remain the core of the space frontier movement. Into the wee morning hours we would argue and plan, cooking up new initiatives and plotting how to change the world in three easy steps. We were creating a conspiracy, a benevolent conspiracy, and our goal was nothing less than changing the nature of human civilization.

Very different from the cheer leading style of other groups, this amalgamation of dreamers had an edge of hard bitten reality — after all, it was our money and time being spent on this mini-space program. We obviously had a different agenda than the space establishment, but they had all the money and power. We wanted to make our side impossible to ignore, in fact we wanted to take over and rewrite the world's space agenda.

For example, in the mid-80's SSI completed the tests of Mass Driver III. I recall standing in the lab with Dr. Les Snively, Dave Brody, Morris Hornik, Dennis Mateik and others as Les ran the countdown. The soft thwack of the small wire coil bouncing off the end of the short tube of coils belied the importance of the project. As Les said then, "We got good data …" An understatement — in fact, the data showed that the device had successfully demonstrated lunar escape velocities, and today almost all lunar and space scenarios include mass drivers in some form. The demonstration inspired some science fiction writers, and got into a few NASA illustrations, but it wasn't enough. Outside of the science and engineering communities, it was essentially ignored.

Despite such breakthrough demonstrations of technologies needed for the human settlement of space, and despite our much more pragmatic, democratic and life centered orientation, NASA and the US space program continued to try and ignore the O'Neillians. The space establishment was still moving down the path laid out by von Braun. Constantly tweaked by the pork-based electoral needs of those in Washington, and the financial needs of the aerospace companies, those in charge of space "just didn't get it." Space wasn't about anything but NASA and satellites to them. Other than running the equivalent of telephone poles through it, the litany of "Station, Shuttle, Moon, Mars" was still completely dominating the US space agenda, and no one was giving the frontier vision more than lip service.

The one bright light of reason and freedom during this period came in 1986 with the publication of the Report of the National Commission on Space, *Pioneering the Space Frontier*. This landmark report, which opened by declaring that "The Solar System is our extended home", was way ahead of its time. Unfortunately, it too was ignored by the space establishment intent on continuing down a failed path.

After the mass driver, came the time to decide what would be next — should the Institute do one big project or go for small breakthrough projects … should it count on what might be a windfall from GeoStar, the satellite global positioning and communications company that Dr. O'Neill founded and was hoping to use as a "terrestrial revenue source" to bootstrap the Institute's work, or should it procede as if there would be no money flowing? Questions were asked about everything, even the location of SSI.

It was decided that SSI would keep working small projects, but would also focus on one major project, the leading candidate at the time being a small probe to orbit the Moon and look for resources. For months, Gerry, Gregg Maryniak, Gay Canough, Jim French, myself and others worked on the problem, looking at ideas, rejecting ideas, and slowly moving towards a plan.

Part of my assigned work was to foment a grass roots effort to fund and fly the probe. We knew it would be a long and uphill climb — after all, America and NASA had zero interest in the little gray world. In fact, many at the agency were negative with respect to the idea, given their ever shrinking budget, and seeing it as a threat to other "more relevant" projects. We were going to be on our own. After all, as ironic as it sounds in a democracy, NASA had never flown a citizen initiated exploration mission.

Citizen initiated. Again, that empowerment thing. Until SSI there was no citizen space program. There were the "right stuff" guys on TV, and the rest of us, sitting on couches watching. From the moment Gerry chose everyday people as his narrators in *The High Frontier*, cracks began to appear in the elitist space program of the cold war. Gerry and SSI said, YOU can be a part of opening the frontier. You can come here to Princeton and share your ideas. You can try and build space equipment for pennies compared to the government dollar, and YOU can even build a space probe.

Well, once you let that "I can do it" Genie out of the bottle ….

I recall several SA's and myself choreographing an event at the Pittsburgh Conference of the NSS wherein I gave a speech that SSI and I had developed. I followed it by pledging 100 dollars or so and asking for a match. Well, I knew I had one, as I had planted it. By the time we were done the first money was flowing to help fund the first citizen initiated space probe project. The *Prospector* team soon turned into a small movement, with allies showing up on the Hill, and support building on many fronts.

We joined forces with Dr. Alan Binder (who later became chief scientist of the mission). Al became the credible bulldog every such project needs, he refused to drop or let go of it until it was flown. He wrangled support from a few enlightened managers at Lockheed Martin (his employer), and eventually *Prospector* was selected by NASA to fly as a *Discovery* mission. The effort also had begun to spin off others projects. For example, the O'Neillian members of a Department of Defence team were able to convince their superiors that the best place to test some of their new *Star Wars* technologies would be the Moon. Thus was born *Clementine*, which gave us the first indications of hydrogen at the Lunar poles, later to be confirmed by the *Prospector*. (*Clementine* even carried a small sub-satellite named *O'Neill*.)

The "you can do it" attitude was spreading. As part of the Lunar *Prospector* effort, SSI held a Lunar Systems Workshop, attended by Freeman Dyson, Gerry, and many others. For four days the team went over ideas for creating profitable enterprises in Cis-Lunar space. Among them was one involving teleoperated vehicles roaming the Lunar surface in a race I called the Millennium Moon Race. Back on Earth audiences would watch the images from the little robotic rovers live, as celebrity drivers strapped into gimbaled rock and roll chairs would maneuver them through a course on the Lunar surface.

The Lunar race in the Lunar Systems study evolved into the LunaCorp business plan when Jim Dunstan, Tom Rogers and David Gump ran with it. In fact, it was at about this time that I left SSI employment to join LunaCorp's effort and run the newly formed Space Frontier Foundation.

SSI constantly throws out seeds of ideas into the space community. It funded the earliest work on External Tanks as habitats, including the seminal study by Alex Gimarc. SSI also publicized efforts by people like Tom Taylor, and External Tanks Corporation, who eventually got an as yet unfulfilled pledge by the Reagan White House to hand over 5 external tanks. This pledge has been dormant, but I think it will be tested very, very soon.

In the area of space solar power, no one has done so much as the SSI family to keep this dream alive ...from demonstrations of power beaming, to funding small, but important work on rectennna technologies, SSI is at the heart of the effort.

The Space Frontier Foundation is a blood child of SSI. Gerry knew of and endorsed our plans. It was the Foundation's belief that in order for space exploration and settlement to take place soon and involve large numbers of people, the message needed to be taken to the people. But SSI was not the right vehicle for this more

The external liquid fuel tank used to launch the Shuttle (which functions for only about 20 minutes) could easily be retained from each launch. The collected external tanks could be used for space construction instead of being wasted.

aggressive assault on the media and political circles. The Institute needed to "stay clean" and remain a solid source of information and credibility for the core ideas we were fighting for.

Working for SSI and leading grass roots efforts for the L5 Society, I had witnessed what did and did not work in the media and politics. I knew we had the best product on Earth — and in the rest of the galaxy. It was time to "kick it up a notch."

The first step was to get real. We had to look at what was happening out there in the world. We had to lose the unrealistic rosy filter of a bunch of sci fi readers "chanting L5 by '95", wherein a pure heart and good science fiction would somehow transform the national cold war space program into a civilian led private sector breakout into the frontier. We had lived that fantasy long enough, we had been there and done that and we could see it wasn't happening. No, what was needed was a long, long view, an understanding that an entire new agenda was needed, that had settlement as its goal rather than as a source of artwork for posters selling projects that had no relevance to our cause. It also meant being willing to ask hard questions, accept hard criticisms, developing a little sophistication and a lot of willingness to play the Machiavellian games of Washington and the corporate world and go for the win.

Unfortunately, we were starting with some serious handicaps, after all, what do you call someone who is out of touch with reality — "spacey" — and this is how we were seen. This was partially our fault; many of our leaders at the time were science fiction writers, and many of our supporters came from that world. In Washington, due to some unfortunate attempts to lobby Congress early in the decade by members of the O'Neill inspired L5 Society (which later merged with the National Space Institute to form the National Space Society), anyone talking about space colonies was assumed to be a flake or *Trekkie*. I was a proud member of L5 and founded the NYC chapter, but the label was hard to shake. And Washington being the understanding town it is, supporters had been characterized by one or two enthusiasts showing up on the Hill in their Star Fleet uniforms. It was clear that if we were to realize the vision, we needed to learn to play the game in Washington, Wall Street, and in the media, and to create legitimacy for ourselves and the frontier ideal in the public mind. In other words, we needed to sell our product more effectively, to change the hearts and minds of the public, the decision makers in Washington and yes, even NASA itself.

Shortly after setting up the New York City L5, the national organization merged with the heavily von Braunian oriented National Space Institute. At stake was a pot of money being held out by the aerospace companies, if the two would work together. Unfortunately, the price was a dilution of the vision, and blanket endorsements of NASA's space station project. Although I had personally asked Gerry to support the idea of human facilities in orbit (and he had done so in writing), buying into the NASA station project with its limited facilities, unclear goals, and its aggressively negative approach to other pro-frontier concepts, was too much.

During this period I had gotten to know Jim Muncy, a Washington insider who was protégé of the young Congressman from Georgia, Newt Gingrich. Although we came from opposite ends of the political spectrum, we shared the vision, and the desire to take this idea "to the streets". Soon Jim and I were working together for SSI, carrying out a set of meetings with Congressmen, such as Jim Torricelli of New Jersey and Dana Rohrabacher of California, to gain support for the Lunar *Prospector* project. Specifically, we were trying to break free an old gamma ray spectrometer that had been scheduled to fly on Apollo 18 and was being stored at NASA's Jet Propulsion Laboratory in Pasadena. Events and technology soon made the need for the device moot, but in the process of working the Hill, I began to realize that with just a little polish and strategic thinking, we could move our agenda ahead, even in the swamps of Washington politics.

At about the same, time Bettie Greber asked me to talk to another long time SSI supporter named Bob Werb, who wanted to get more involved, and made the mistake of driving me home in his pick-up after an SSI and NYC L5 event in 1986. By the time we arrived at my apartment, he had agreed to lead the printing of a new edition of *The High Frontier*, and I knew I had found another partner. Bob came from yet a third, very different background from Jim and I, being a serious real estate developer and practicing free enterpriser with a strong liberal bias. I figured that, if the three of us, from such different backgrounds, could unite under

Gerry's vision and reach consensus, then our public case would be much stronger. This philosophy of strong, committed and wildly different types of people agreeing to knock-down, drag-out fights in the back room and then introducing their fire tested ideas to the public in a strong and unified front, was to become one of the keys to our success.

After several intensive meetings, Jim, Bob and I began to create a plan. It was decided that we needed an action oriented organization built around projects that would give what Bob called "serious professional volunteers" an umbrella and support network so they could do their best work.

We liked the SSI Senior Associates model, wherein an "elite" core of leaders worked closely together, almost as a family. We took this idea a step further, and gave this group, whom we eventually and after much wrangling decided to call "Advocates", the power to elect the Board and make other substantive decisions.

Knowing the external forces we would face, we didn't want to become a debating society and tear ourselves apart as so many groups do, so we set up a doctrine, and designed methods of screening out those who didn't share our views. In fact, we saw ourselves less as a club or organization than an army in the field of ideas. And we definitely did not want to be "yet another space group."

As far as we were concerned, if you had been invited and made it through the process, we assumed you already knew the concepts we were fighting for, and that you knew what you were doing. Yes, we have a belief about the future, and we do stick to it. Our goal is to have the world move towards the O'Neillian vision, for we believe this is about nothing less than species survival and prosperity. We have a high sounding vision statement that summarizes that idea, but then we roll up our sleeves and focus on what needs to be done today to move us forward. Sometimes this means loud public fights with the establishment; sometimes it means quiet behind-the-scenes legwork; but our commitment is to effectiveness, not useless action or bellicosity. In fact, that focus led one space leader to call us "pound-for-pound, the most effective space organization in America."

However, we knew we often underestimated the sophistication sometimes demonstrated by the bureaucratic beast we were taking on. We had seen good people and ideas out maneuvered and/or eliminated time and time again. We wanted to be very aggressive, but also not so shrill as to be dismissed, so we studied and applied the lessons of movements and media in the past.

We decided to carry on simultaneously what was called "the inside game" in Washington and the space community, and the "outside game" in the media and public. We knew we couldn't simply "tell the truth about space and the space program" as Muncy put it, we had to make it credible. We had to be as slick as Madison Avenue, yet as credible as the *Wall Street Journal*. We decided to "take a stand" for the frontier, and hold that stand until we won.

Our first project was chosen with care. We knew we had an Apollo Anniversary coming that we could use as a hook — taking a celebration of the past and spinning it with our version of the future. We also knew that the opening project would determine the kind of people we attracted. And of course, the results and style of the activity would be scrutinized by the space community and media, as they gave us their first "sniff" test. Thus we went with a national petition drive that called for a commercial and scientific "Return to the Moon."

Although we were beginners, we were trying to use the tools of the media as well as we could. The concepts of PR and propaganda were something we wanted to apply to the cause, and we did so with a vengeance. We created a category of slogan we called "cultural cruise missiles", based on the idea of codifying a concept in a slogan that would then be launched into the culture to seek its own targets. The terms "Return to the Moon," "No More Flags and Footprints," and "Next Time We Stay" were peppered throughout our literature.

Within a few months, under the incredibly dedicated leadership of Kevin Griffin and Elisa Wynn, over 50,000 signatures calling for a commercial and scientific lunar base were on the desk of then President George Bush.

Our slogans of "no more flags and footprints" and "next time we stay" were incorporated into his 20th Apollo anniversary speech. Now, in other areas, such as social and environmental movements, this would be seen as a small drive, but for space it was a first. Its only equivalent had been a PR oriented drive in the late 70's to give a NASA Space Shuttle the name *Enterprise* in honor of *Star Trek*.

At our Space Frontier Conferences we give those politically motivated within the space field a gathering place. Our conferences are aimed at what we call "players" in the space field. Those who are partnering theory and research with venture capitalists and entrepreneurs to make it all happen.

The DC-X (shown here during its initial hover test) was designed as a reusable surface-to-orbit vehicle. The DC-X was the predecessor to the X-33.

As part of our outside game to influence the popular culture and reward true heroes of the frontier, we created a set of awards for *Best Vision of the Future* and the *Vision to Reality*. *Vision to Reality* is only given to those who have succeeded with a space project. The winners include the DC-X Re-Usable Demonstrator team, Space Hab for their work in commercial space services, *Clementine* and the Lunar *Prospector* Teams. *Best Vision of the Future* goes to those who have done the best job presenting our future in space, and includes such TV shows as *Babylon 5* and *Deep Space Nine*. The idea, to bring Hollywood and the space field together in one room, create some chemistry and let the cross fertilization begin, spawned the Los Angeles conference. The Hollywood community, with more than sufficient backing, knows how to put on a good show, and is always looking for something new. The space revolution is new, has things to teach Hollywood about real space, and needs the kind of positive PR and capitalism that Hollywood is famous for. The concept seems to be working, and we're slowly creating a serious space aware community in Hollywood. All part of our re-education effort, all tied together.

Moving from symbolic petitions to serious political action came fairly easily to the Foundation, as we realized that the only way to get what we wanted in space was to follow the money and the power. That meant Washington and Wall Street. And the biggest concentration of both in space was Reagan's multi-billion dollar space station project.

When it came to what was then called space station *Freedom*, the problem we had to confront was ironic. To us the station program's structure assured minimal participation, and killed hopes for a real frontier to develop. We didn't want to see the Antarctica model (a few national and international outposts in a heavily government controlled domain) repeated in space. We believed that if the need was really there for such a facility in orbit, the federal government could simply characterize the need and lay out the financial and legal groundwork, and the private sector would provide one or more such facilities. Otherwise, the turf building character of the agency would assure its own hegemony in LEO for as far into the future as we could see. Yet the division between pro- and anti-space station activists meant that we, whose whole motivation was the human settlement of space, found ourselves for several years opposing the first permanent US outpost in space, the only pro-humans in space organization to do so.

The space station project was not only going off the rails, but was eating its young, in the form of private ventures that although they were spun off of the NASA plan, were seen as threats by the bureaucrats. We fought for new alternative ideas such as External Tank based facilities and helped win an order from President Reagan to hand over a few tanks to private firms who could take them on as government surplus. We moved our headquarters into the mouth of the beast, setting up literally across the street from NASA's Johnson Space Center. At the same time we jousted with NASA Administrator Truly in *The New York Times* over the state of the station project, and even found ourselves receiving personal threats over our positions on the multi-billion dollar fiasco.

We knew we needed both destinations and transportation, so we were growing an effort based around a term we created, called "Cheap Access to Space" (CATS). In this case, we got to be "for" something, and it was frankly refreshing. It had become clear to us that massive government projects were not helping open the frontier, and the cost of space transportation was sealing the door shut. This situation was due to complete dominance of the market by a government using ballistic missile technology to fulfill its own case by case needs, with no eye to creating transportation systems in the commercial sense.

The frontier movement needed a symbolic Conestoga. And we got one. With the Air Force's acceptance of a plan to build a small vehicle called the DC-X that was designed to show extremely low cost operations, we knew we had a flagship for our cause. Ironically, this project forced the supporters of O'Neill to work with those who came from the military side of space. In fact, it was O'Neillians and supporters of the Space Defense Initiative who drove the effort. NASA was still promoting expendable solutions, and Dan Goldin, the NASA Administrator, was pooh-poohing the new single stage to orbit concepts just weeks before the DC-X Project was announced.

Soon, working with other groups, we had become the leaders in the fight for CATS, trumpeting the paradigm breaking qualities of the "little rocket that could" and educating the media and policy experts as to the significance of the project. During this time we developed our Internet strategies, becoming early users of such things as email alerts and distribution lists. Volunteers were assigned to literally nag reporters on our perspectives until they submitted, and others began to work in DC on political strategies.

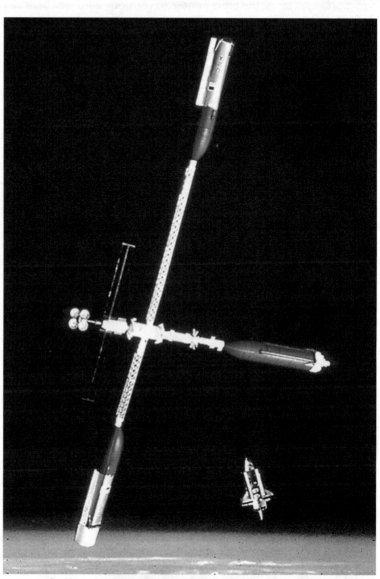

In this proposed scenario, Shuttle External Tanks would be used as the primary raw material for building a gravity-by-rotation test station to further knowledge of rotational gravity's effects on human physiology.

We reached the capability to place an article in the press, send out a release on the net, alert a mass of write-and-call-in volunteers and place an item on the Congressional agenda in a coordinated attack. All of our press releases and op-eds on these issues mentioned the others, to get readers to begin to understand the interrelatedness of these various frontier projects.

By 1992 we had helped station opponents get within one vote of killing the program. At the same moment in time, White Sands had literally locked the now broke DC-X team out of their test range, and the Air Force was going to put in no more money. We knew we needed the DC-X badly to win over the public. We needed the four minute mile of a spaceship that could take off and return in one piece, fly often and fly cheap, splashed across the TV screens of the nation and the world.

At the point of that oh so close vote in Congress to kill the project, we knew we couldn't win. Our argument was good, it had historical precedent, it was right. A single government owned and operated outpost versus creating the means for the participation of Americans of all sorts in opening the grandest frontier of all time, and most importantly, the press printed us. But our third way, between the NASA approach and the luddites "why waste money on space idealist fantasies," was simply lost. You were for space station or against it. No third parties. The very valid concepts for commercial free flyers, and the huge potential of recycled External Tank based facilities, were lost in the noise, or dismissed by the NASA establishment of the time. After all, if these were such good ideas why wasn't NASA pursuing them?

We were too small and spread too thin. Our volunteers were literally burning out around us, with some even leaving the field completely to try and regain their lives. We realized that we had a choice, we could continue to fight a losing battle over the station, constantly on the attack, or we could turn our attention to a positive fight that would help open space to all, and if the station survived, return to it later.

Using channels that we had developed into the top level of the space agency, we made it known that we and our allies might consider backing off our station attacks, if NASA would move in and support the DC-X project. Although we were still only a small organization, our ability to place our bold media messages (usually stripped of our positive content it seemed) and our assistance to others was dogging the program's supporters. After all, we weren't mere anti-technology luddites, we were from the space community. We could walk the walk and talk the talk. And, when major votes are passing by a hair, small groups can wield influence far beyond their size.

Thus, later that day we got a call saying several hundred thousand dollars was being transferred to the DC-X project and NASA was moving in. DC-X was saved. It was a turning point in the battle for CATS, but the future would hold even greater challenges.

The X-33 reusable vehicle grew out of the DC-X project (Cheap Access To Space) once NASA became involved.

During the CATS fight, we had developed a lot of savvy regarding politics and space. Jim Muncy and Tim Kyger were our insiders in the House and Senate. They were finding new allies in Congress, among them former Congressional Whip Robert Walker, and Chairman of the House Space Subcommittee, Dana Rohrabacher, who had helped on the *Prospector* project.

With this help, and the credibility we were gaining in the media, we carried the CATS fight forward on many fronts. By the following year we had gained White House and Congressional support for the development of the X-33 reusable space plane test vehicle. Although our goal was to see two vehicles built by different firms competing in the process, we ended up with one, and an alternative called the X-34. Today it looks as if the X-34 may well develop into a useful system, while the X-33 withers on the vine, the victim of over promising, bait and switch, and simple corporate indecision.

We did move on in other fronts, greatly shifting our focus away from government efforts and towards the work being done by the new crazy guys and their rocket machines — a group of small and diverse space transportation firms born at the end of the cold war to fill the need of placing hundreds of new satellites in orbit, and to dream of carrying citizens into space. We ran the first ever CATS event in Washington, ironically underwritten by a 100k grant from NASA. This was a very mainstream type event and allowed us to put Gary Hudson of Rotary Rocket and Pete Conrad of United Space Lines on the same platform as Lockheed,

Boeing and the FAA. The idea was to make it clear that there was a new game in town called competition and free enterprise. We have yet to see if the game will be played and won.

Currently, we have helped forge the space transportation companies into a recognizable industry, dedicating a full day of each of our conferences to them, and promoting concepts such as tax credits and payload purchases by the government, much like was done to help kick-start the airlines in the 1930's.

During our fight for CATS we had noticed a young guy named Chaz Miller, who had worked with us, had been fighting for a private Lunar base, and was

The X-34 reusable vehicle is the successor to the X-33.

also organizing events on the Hill, so we brought him in. With his help we spun off our political activities into a new lobby called ProSpace. In the first year, a few dozen volunteers paid their own way to Washington and spent a week working staffers across the Hill. Their methods were the most sophisticated ever seen in the space field, involving pre-training, simulated meetings, computerized scheduling, tracking of results and intense follow-through after the event. Included in the week long annual project were high level receptions for members of Congress and events in the House attended by the Speaker and others.

ProSpace has had a large effect on space policy, as long term relationships with staffers were built and a track record was established. For example, recently it was ProSpace that worked with other groups such as the NSS to drive the *Space Commercialization Act* to its final passage. This piece of legislation was drafted by David Anderman, the first Foundation executive director, among others. ProSpace recently drove the allocation of several millions of dollars towards Space Solar Power (SSP) development, something that was barely in the works before our team arrived on the scene. This effort is continuing today as the Foundation, ProSpace and our friends on the Hill begin to expand this wedge of work into a real SSP R&D effort.

Being pragmatists as well as zealots, the Space Station wars forced us to re-examine our strategy. So a few years ago the Foundation adopted a new psychological and strategic model for our battle, based on the martial arts concept of jiu-jitsu. The idea behind the concept is to use the momentum of an overwhelming opponent's charge by redirecting it slightly, and using that force to gain your own advantage.

Armed with this new mind set, I began to take a new look at some of the US's most ambitious and seemingly unstoppable projects, such as the International Space Station, the Space Shuttles, and the almost religious drive in the agency to someday send humans to Mars.

It was time to redefine space for NASA, the media and Wall Street. For decades it had been the mismatch between ability to deliver and expectations that had derailed so many good ideas in space. The "NASA will open space for us" crowd was wrong, as NASA was simply not going to build Hiltons and space liner fleets a la Arthur C. Clarke's *2001*. The "get the government out of it all the way" camp was also wrong. The private sector was neither able nor willing to simply jump in and throw open the frontier without some participation and support from government in the early stages. Space is seen as so exotic, so dangerous and so difficult to work in, and NASA has become so dominant in our culture, that people couldn't make the leap from *Star Trek* to real possibility. There was no mental road map. And without something to make the High Frontier real, without "permission to dream" from the institutions and creators of thought in our culture, no power company manager was simply going to look up into space and say to herself "Space solar power is the solution!"

The idea was obvious. If NASA is our modern day Lewis and Clarke, then they should get out to the edge and do what explorers do, go over the hill and report back to us what they find. When it comes to driving Space Shuttle trucks and constructing buildings, the private sector takes over, creating new human societies.

NASA found itself somewhat trapped by the Station and Shuttle, as these two projects began to eat its budgets for science and exploration. The three hundred pound space gorilla had found its hands full with projects it couldn't eat, squatted down in low Earth orbit, and wouldn't let anyone else into its cage. We ramped up our efforts to point out that neither job was the right role for the government, combined with strong hints that if the agency could extricate itself from Station and Shuttle, they could then begin to look towards Mars, NASA's promised land.

I tried to make this concept concrete with Alpha Town. The goal being to transform the International Space Station into a nexus for commercial activities, using the staying power and cash flow of the federal anchor tenants to catalyze the development of a private sector economy. Run by a Port Authority or consortium, the Space Station would be much like an industrial park, playing host to a wide variety of activities, with the government agencies that built her being the prime renters. These tenants would work through the management firm to purchase their needs from vendors. Alpha Town called for NASA and the ISS operators to purchase all energy from private vendors off site (SSP), buy transportation from a wide variety of commercial bidders (CATS), and lease its housing needs (ETs, Mir etc.) from the private sector. Eventually, I believe the synergies and new relationships this arrangement will create will transform how we think about space, setting new precedents that may last centuries. Servicing the needs of this facility, and new ones that may spring up near by, will lead to a community springing up around the Station, just outside the airlocks, so to speak.

These ideas are catching on even as I write this, as the Station moves inexorably towards the Alpha Town model. New forces are arising that may lead to a pick up in the speed of our opening of the High Frontier. Wall Street is slowly becoming aware of the new industries space may deliver and is watching with wary expectation as it ever so slowly matures.

Another new force, long awaited by the stalwarts of the frontier faith, are the knights of commerce, those few visionaries with the means to succeed independently. At the beginning of the 21st century, we're seeing the rise of the first generation of incredibly wealthy entrepreneurs and dreamers who were raised on space. Coming from the communications, entertainment and service sectors, these new barons of industry have a lot of money, care about the planet and grew up watching *Star Trek*. Some of these people have begun to see the flickering candles held up on the frontier by the followers of O'Neill, and they are beginning to get involved. The wonderful thing about this development is that it's based on individuals, not boardrooms or bureaucrats. One not so crazy billionaire told me that, by himself, he can put people in space, make a profit, and to hell with the paperwork! Now that's a frontier attitude!

There are major hotel related fortunes being applied to the development of space hotels, with employees being hired and factories being built. Another would-be space entrepreneur is well on the way to building a totally commercial rocket capable of carrying large payloads for a fraction of today's costs. Others are getting involved all the time. These space cowboys are a totally new element on the critical path to space settlement. Frontiers attract people like these, and the confluence of our "Space is a Frontier" message, the real data we have now that shows space can be opened effectively, and the economic boom of the last decade make it the right time.

Nearer to home, one of these new space patrons was convinced by me to put a few million dollars into an endowment aimed at supporting what SSI calls "critical path" work for the frontier. At last, the frontier movement had the fund that Gerry had tried to create with profits from GeoStar and other ventures.

Called FINDS (the Foundation for the International Non-governmental Development of Space) the endowment now stands at almost $30 million dollars. Designed to literally "Find" cutting edge non-government space projects related to the human settlement of space, FINDS joins SSI as a provider and supporter of space research and development. Due to a minimal structure and a strong network of friends and information providers, FINDS is able to move quickly and aim small amounts of money, just when needed, to leading edge projects.

To date FINDS has funded a wide range of pro-frontier work. We are supporting the search for Near-Earth Objects through a joint project with the Space Frontier Foundation called the WATCH, and we're developing actual asteroid processing hardware with the Universities of Arizona and Woolagong, Australia. FINDS is working with SSI and the Russian firm Energia to study the layering of solar cell and rectenna materials on thin film for solar power generation. We are also funding a solar sail project and lunar ice hunting robots with Carnegie Mellon University, helping SETI search for extraterrestrials, and supporting researchers and students in several countries around the world. With the Space Frontier Foundation, we are offering the first fully funded space prize (the CATS Prize) for the first non-governmental team to place 2 Kg at an altitude of 200 km. In house, we're working on a large project with the Russians to examine the use of tether systems as a low cost means of orbital control for large space structures. FINDS also supports events like the SSI Conferences by offering prizes for papers, and giving funding to symposia and studies such as the first Lunar Base Symposiums, and other frontier advancing events.

The relationship between SSI, the Space Frontier Foundation, ProSpace and FINDS continues to be strong and interactive. Since George Friedman became SSI's research director, he has stressed the increased importance of near-Earth asteroids, both as a threat and, more importantly, as a source of extraterrestrial material far more diverse and accessible than the lunar surface. As I have just summarized, we are vigorously implementing these new directions which were not obvious to Gerry, nor indeed the entire astronautical community, during the high frontier's early days. On the other hand, George has been receptive to my concepts of space habitat construction. These involve building of relatively comfortable enclosures around the construction site that permit a shirtsleeve, rather than space suit, environment for space workers. This would substantially increase their safety and efficiency. Additionally, George, as well as Freeman Dyson and John Lewis are Foundation advisors, and there is a healthy interchange between the organizations as they support each other's conferences. Indeed, the Foundation and SSI are cosponsoring a Lunar Development Conference in Las Vegas in July, 2000.

The space frontier movement is alive, and can now carry the title "movement," as there is a revolution underway. Gerry may not have been happy at how slow things are progressing, none of us are. But he was a realist at heart, and knew our work would not be easy. However, I do think he would be pleased to see that the revolution he began is building over time, and that the ideas he planted are taking root in the hearts and minds of new generations. Thanks to the hard work of ProSpace, legislation is finally being put together in the US to support the opening of the frontier through tax and investment incentives. The Foundation is slowly changing the image of space presented by the media, educating reporters and Wall Street as to the potential of the frontier and working with possible new allies such as environmental groups to show that we are on the same side. Finally, FINDS and SSI are providing the hard data and examples of what can be done now to open the path to space for all. The near frontier is transitioning to the private sector ever so slowly, as we fight for privatization of the ISS and space transportation systems. On the far frontier, even NASA's plans for Mars have been infected with two of the prime hallmarks of the O'Neillian conspiracy, permanence and the use of in-situ resources to make that permanence supportable.

In a few years there will be a bustle of human life in the near frontier of the Earth and Moon. I believe humans will be engaged in the full scope of activities we find on terrestrial frontiers, from science to entertainment, exploration to tourism, creating breakthrough products of all kinds and discovering new ways to live together as new societies are born. Out on the edges we will see humans begin to push back the darkness of the far frontier, as Mars sees its first permanent outposts established, and asteroids from Phobos and Deimos out to the main asteroid belt feel the touch of humankind.

I firmly believe that by the middle of the next century, we may well be part of a culture that sees itself at the center of an ever expanding bubble of life, rippling outwards from its womb on planet Earth. It will be clear to any child born inside that bubble that destiny lies on its edge, and that existence is defined in terms of life's expansion outwards.

Gerry O'Neill's vision of the High Frontier was never about tin cans in space, or a particular technology or magic economic bullet. It was about life, people and dreams. He pointed to our hands and said use them, he pointed to the horizon and said go there and learn how to live, he pointed to our hearts and said feel the strength of what it means to be human. He challenged us to take on a universe that can wipe us out in a moment and say, "We are here! We live! And we are coming!"

The High Frontier is alive.

# Chapter 18
# At Millennium's Eve: The View From 1999
by
George Friedman

We are caught in the midst of a complex puzzle. Let us not delude ourselves into thinking that the solutions will be simple or that others will do our work.

From the perspective of the eve of the new millennium, the primary elements of Gerry O'Neill's vision for humanity and space have remained a bright and clear beacon of hope over the two decades since the first publication of *The High Frontier*. His observations, as viewed in the 70's, are still dramatically insightful and prophetic today.

Although Earth is still a paradise, humanity is in a trap at the bottom of a gravity well.

Our population is increasing by a billion souls every dozen years and even the most optimistic demographers admit that the total will exceed ten billion early in the 21st century. The greatest increases are in the underdeveloped nations, who have aspirations of substantially increasing their standards of living as well. The impact on Earth's biosphere will be beyond anything mankind has ever witnessed — not only the well publicized degradation of our fragile environment, but the extinction of more species than were killed after the cosmic impact that ended the great age of the dinosaurs and formed the Cretaceous / Tertiary boundary 65 million years ago.

To avoid this dire future, humanity's available resources and wealth must increase at least as rapidly as the population increases. Energy is perhaps the single most crucial resource both in supporting life and in determining the quality of life. Even if there were no negative aspects to the burning of fossil fuels, we're burning them up a million times more rapidly than they were created.

In direct contrast to the "limits to growth" philosophy that proposed a static society based on zero population growth, Gerry O'Neill proposed a high frontier which not only opened up virtually limitless sources of energy and material resources, but also had deeply positive benefits to humanity's psychology. Employing only existing technology and straightforward application of engineering principles, O'Neill demonstrated that breaking the artificial "limits to growth" barrier was possible *today*, requiring only the allocation of resources, not a technological breakthrough. Space is clearly the path for a race with infinite aspirations but finite resources.

One of the earliest exploitations of space should be the building of a constellation of Solar Power Satellites with the capability of collecting Sunlight in orbit, converting the energy to microwaves and beaming the power to receivers on Earth at an overall efficiency an order of magnitude higher than terrestrial solar power. This concept is by far the most robust approach to attaining the growth of wealth at a rate equivalent to population growth. The progress of this concept is reviewed in chapter 15.

As humanity requires additional room to live, colonization of space should occur — not on the surface of Mars or the Moon as envisioned by science fiction and the Mars Society — but within space habitats. Mars is a thousand times less habitable than Antarctica, which is a thousand times less habitable than Siberia, which is a thousand times less habitable than Hawaii, which is representative of the climate attainable in a space habitat. Moreover, while space habitats can easily provide simulated gravity by means of rotation, migration to the surface of any planetary body in the solar system would be a one way trip — return to Earth would be extremely difficult because of the differences in gravitation and the long term adverse effects of living in a low G environment. Even with these disadvantages, at the most optimistic, Mars could possibly support a tenth of humanity, whereas the exploitation of space resources as explained in chapter 16 could eventually support a million times our present human population.

O'Neill stated that the schedule for these advances is not dependent upon science, "but a complicated, unpredictable interplay of current events, politics, individual personalities, technology and chance." Even having said that, the actual schedule is far slower than even his most pessimistic projections for achieving occupied space habitats. Why is it taking so long? The L5 society — inspired by

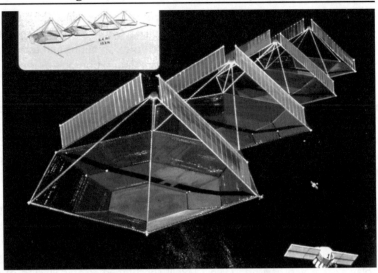

One of the earliest exploitations of space should be the building of a constellation of Solar Power Satellites.

O'Neill's vision — had a slogan: "L5 in '95!" They certainly didn't mean 2095. Most importantly, is humanity even on a path that will achieve the beginnings of space exploitation and colonization by 2095? These are the most crucial questions for the advocates of space colonization and the remainder of this chapter will be devoted to their discussion.

Speaking as a lifelong space enthusiast, I must confess that our group of idealists has been subjected to — and perhaps has succumbed to — at least five tempting illusions:

First, there is the illusion that the US space policy and funding during the Sputnik to Apollo era can be extrapolated into an ever increasing and ambitious program of science, exploration, exploitation and colonization. In reality, NASA's primary mission was to help win the cold war by countering the temporary Soviet space leadership in the 1950's and early 1960's. The cold war is now over and the priority for space and defense is diminished. Advocates for feeding, educating and medicating the poor, and for protecting the environment are competing more successfully for funding. They suggest that space can wait — as they have patiently waited — until problems of more immediate relevance to society's needs are solved. Instead of competing with the Soviets, we are now propping up a collapsing Russian space program on a space station of questionable value and enormous cost. Freeman Dyson observed in the Introduction that there is a real NASA in charge of programs and a paper NASA looking so far into the future that the concepts are rarely realized. Perhaps even more damaging to our cause has been NASA's policy of managing all operations in such a manner that entrepreneurial private investment was stifled. The goal of reducing the cost to lift mass from the Earth to low-Earth orbit by orders of magnitude from its current ten thousand dollars per pound has been a complete failure.

Second, there is the illusion that manned missions to space are analogous to the exploration and colonization of the western hemisphere. Five centuries ago, explorers found the fertile land, fresh air and clean water of a literal paradise in the western hemisphere, with an abundance of edible plant and animal life, even including a populace so closely related that cross breeding was possible. By contrast, the tiniest rip in a spaceman's suit threatens a horrible death, and even in the best of cases each person in space must be supported by hundreds of tons of protective structure and equipment. Although we correctly claim that the resources of space are virtually infinite, these resources are incredibly diffused within gigantic volumes and will take careful, patient engineering and pragmatic sequencing of business opportunities.

Third, there is the illusion of rationality; that is, if a new technology advancing the state of the art is made available, it will be used, and if a new strategy solving a potential problem is formulated, it will be implemented. Most decisions are not made on a strictly technological or decision / theoretic basis, but are rather based on a multitude of diverse reasons which include timing, market forces, competition, public opinion, greed and personality. An example is Gerry O'Neill's Geostar concept for global navigation and communication. An honest comparison with the competing and established Global Positioning System (GPS) — which was originally conceived for military application only — would surely show that Geostar is simpler, cheaper and provides a communication channel that GPS does not. However, the early presence in the marketplace, coupled with the ability to take advantage of miniaturized and inexpensive electronics, permitted GPS to virtually suffocate Geostar in its cradle. A hoped-for stable financial base for SSI thereby was dissipated. Another more relevant example is that, despite O'Neill's excellent list of the advantages of space habitats over planetary surfaces for space colonization, Mars — due to some combination of the visions of Percival Lowell, H.G. Wells, Orson Welles and Robert Zubrin — continues to capture the attention of the masses, while space habitats still occupy only a small corner of the public consciousness.

Fourth, there is the illusion that, following a crucial catalytic "ignition," growth will be exponential. A NASA summer study in 1980 strongly recommended the development of self-replicating systems enabling factories which would reproduce themselves and eventually consume entire astronomical bodies. It can be shown mathematically that doubling times of a few years in space operations yield enormous expansions into the solar system and beyond. Even O'Neill succumbed to this illusion of extrapolation as he talked about an amplification of humanity's capability by a factor of $10^{88}$. Even a factor of "only" $10^1$ should be sufficient to assure humanity's survival, and I would be more than delighted if it took centuries to achieve a factor of $10^3$. What is ignored by those who indulge in the sterile arguments of whether the exponential or Fibonacci series is more appropriate is that two events are essential for exponential expansion — a robust start to the process and ample resources *of the proper type*. Despite the many references to John von Neumann's "kinematic model" of the 1950's, we still are not even past the conceptual stage in self-replication, and as was mentioned earlier, we certainly have not achieved a start of any sort in breaking out into space with traditional technology. Regarding adequate resources, despite our claims that space has "virtually limitless" energy and material, it is presently not in the proper form or concentration for immediate exploitation. von Neumann assumed that his self-replicating automaton would be immersed in a "sea of parts" — it wouldn't have to be autotrophic and able to live on elemental substances, as we must do in space. Even on Earth, with a fully operational biological self-replicating system and a vast array of fertilizers with which to coddle our crops, our farmers have a hard time producing slight yearly increases in their yield, much less exponential growth. Only over the perspective of thousands of years of agriculture does our food supply appear exponential. In short, the difficulty of transforming limitless energy and material into useful form has been overlooked and sometimes lost in the glare of a fascinating but irrelevant mathematical fantasy.

Finally, there is the fifth illusion that science fiction (SF) is an accurate predictor of technology and social trends. Although the fascination of science fiction has inspired the imaginations of untold thousands of engineers and scientists — and reportedly was influential in their choice of careers — SF's primary purpose is to entertain and is necessarily superficial in order to reach the broadest audience. Despite a few notable successes (submarines, atomic power, air travel . . .), SF has not been an accurate predictor of new technology and its applications. Rather, most stories *assume* a scenario based on new technology and then develop the human consequences. Most people cannot distinguish between serious philosophers of science such as Bernal, Tsiolkowsky, O'Neill, Dyson and Lewis on the one hand and the authors of the motion pictures *Deep Impact* and *Armageddon* on the other. In fact, relationships with SF can have a negative effect, especially in the minds of government decision makers. One reason that the study of a planetary defense system to protect the Earth against collisions by large asteroids and comets has experienced funding difficulty is that few members of congress or the executive branch want to be associated with irresponsible decisions based on a hazard which had its origins in SF. The two motion pictures mentioned above, although raising the public consciousness about such a threat, served to deepen the illusion that this potential threat is merely imaginative fiction.

The previous five rather lengthy paragraphs attempt to explain in retrospect why the pace so far has been disappointing with respect to our earlier hopes. These observations are certainly not a cause for despair. On the contrary, they can provide the focus for future work and there are several activities and trends that promise to have positive impacts on the objectives of the high frontier.

Perhaps the most noteworthy event of all — and hardly mentioned anymore — is that the cold war is over, and we are infinitely better off than if the projected thermonuclear exchange with the Soviet Union had occurred. Despite the apparently never ending worldwide minor conflicts, we now have massively more resources available for the betterment of mankind than if the cold war had continued or tragically escalated into a hot war. We merely have to make our case more convincingly to a broader constituency than the idealistic space enthusiast sector receiving most of our previous attention.

Accordingly, over the past few years SSI has deepened its relationship with two of its offshoot organizations, the Space Frontier Foundation and ProSpace. As summarized by Rick Tumlinson in chapter 17, these organizations broaden the scope of advocacy for the high frontier to the political and public relations sectors. The memberships of these groups tends to be less technical, much younger and substantially more vocal than the average member of SSI. Members of SSI sit on SFF's board of advisors and there is increasing cross-representation at the Princeton SSI, Los Angeles SFF and Washington ProSpace conferences.

In March of 1998, I had the pleasure of participating in ProSpace's "March Storm" which visited the offices of over 400 members of the US House of Representatives and Senate. The week, although exhausting, was absolutely invigorating and even inspirational. Despite early concerns that congressional members and their staffs would be aloof to the ideas of private citizens and that their short attention spans are dominated by near term social priorities, they were invariably courteous and open minded to new ideas. *Most of what we were advocating was passed!*

At the top of ProSpace's priority list was the passage of *The Commercial Space Act of 1998*. This crucial act was passed by both houses of Congress in October 1998. Rep. James Sensenbrenner (R-WI) stated that:

"The growth of the commercial space industry has been limited by government regulations. Historically, we're used to governments dominating space activity. This act will provide policy for the federal government to promote a stable business environment for the commercial space industry."

Rep. Dana Rohrabacher (R-CA) described this act as the "turning point in US space history." Hopefully, NASA's future relationship to space development and operations will follow the excellent example that has occurred for Earth orbit development, wherein NASA provides the high risk technology and operational basis for communications, navigation and observational satellites, then turns the operations over to private industry, allowing an eventual trillion dollar annual business to develop.

ProSpace also strongly advocated an increase in the budget for conceptual studies of Space Solar Power satellites. The budget for these studies was tripled to $15 million, which should give this key high frontier concept a new breath of life, considering the negative results of the early 1980's. Space Solar Power has captured the attention of the National Science Foundation and a special conference on this subject was held in April, 2000.

A crucial program for SFF and ProSpace is Cheap Access to Space (CATS). As Freeman Dyson explained in the introduction, the extremely high cost of lifting mass into orbit is the primary barrier to the economic feasibility of virtually all space ventures. Dyson's "highway" approach, which goes beyond rocketry and doesn't require the lifting of fuel to be expended later, is being stimulated by SSI. These concepts include the mass driver, the slingatron and beamed power propulsion.

The technological disappointment of failing to lower the cost to orbit over the past few decades is partially compensated for by the almost miraculous windfall of low cost electronics and miniaturization. The cost reductions of 10x and 100x in dollars per pound into orbit didn't happen, but size and cost reductions of 1,000x and more did occur in the field of electronics, computers and automation. The feasibility of extending these largely commercial technologies to space telerobotics was the subject of an SSI sponsored PhD dissertation at USC in 1998. Another SSI stimulated PhD research program was the even more aggressive application of molecular nanotechnology to space operations. Of course, not all space activity can be automated, but where it can, every pound of payload can benefit from an enhancement of effectiveness by a factor of hundreds or thousands by applying this advancing technology of miniaturization. According to the conclusions of oversight panels as far back as 1980, NASA's "right stuff" strategy was not sufficiently open minded to permit the replacement or sharing of space operations tasks between man and machine.

As recently as 1988, Gerry O'Neill stated:

> "...the nearest source of materials is the Moon, and the next nearest, typically 1,000 times as far away, is the group of Earth approaching asteroids."

Over the dozen years since then, this notion of the emptiness of the solar system has been overtaken by the relatively enormous asteroidal population predicted by the latest models of planetesimals, formulated by Gene Shoemaker and others. For example, these models predict that there exist over 100 million asteroids with diameters from 10 meters to 10 kilometers whose orbits cross the orbit of the Earth. This represents 60 trillion cubic meters of potential space based material resources, much of which contains a far more diverse variety of material than the surface of the Moon or Mars (see chapter 16) and requires far less delta velocity to exploit. The escape velocities of even the largest of these is trivial compared to the Moon's. These objects are truly at the top of the gravity wells of all planets and moons of our solar system — the only gravity well they're in is that of the Sun. (Let's postpone worrying about the gravity well of the galaxy for a while.)

This new information suggests a respectful reconsideration of O'Neill's strategy of using the Moon as the primary source of space material and employing mass drivers to send pellets of material to L5, where space habitats will be constructed. Subsequent analysis by SSI concluded that there was no compelling advantage to L5 over other high orbits. If not L5, then where? Well, why not go to where the mass already exists in large diversity instead of patiently transporting it from the Moon with its limited diversity? Consulting the planetesimal model again informs us that there are 100,000 Earth crossing asteroids with diameters of 120 meters or more. Each of these can provide the mass equivalent of the *Island One* space habitat: 4 million tons of structure and shielding, 400m in diameter with a 2m thick shell, providing a safe, uncrowded, pleasant habitat for 10,000 people. We merely visit one of these newly found neighbors in near-Earth space and "unpeel" it to form over 99% of the mass of the *Island One* design. As before, we'll still need to import the "vitamin parts" — mainly crucial trace elements, water, electronics and biological material — as we strive for a self-sustaining space colony. The period of these asteroid orbits will typically be two years, approaching the Sun between the orbits of Venus and Earth and reaching out beyond Mars to the main asteroid belt at the farthest. The design and control of the habitat mirrors can create a far more stable internal environment than what we experience by the seasonal variations on Earth. Realizing that there will be massive, wide band communication and entertainment opportunities between Earth and each colony, and between colonies, my personal opinion is that all desire for interacting with "real people" can be satisfied by the population of 10,000. For those who are unsatisfied with a small town and still yearn for the life of a city, reference once more to the planetesimal model tells us that there are 7,000 Earth crossing asteroids with diameters over 400 meters, providing the mass to build colonies supporting a population of 360,000 each. Thus, for these two simplistic choices, existing asteroidal resources can provide the foundation for one billion space colonists living in *Island One* towns and an additional two billion living in small cities. Even an order of magnitude lower population should suffice to accomplish one of O'Neill's primary objectives: that of assuring the survival of the human race against any conceivable terrestrial or cosmic calamity. As is described below, mass drivers will still be needed for a great variety of transportation needs, but the crucial point is that we can build the habitats where we find the mass and don't need to transport it from the Moon expensively and laboriously.

Scratching out a few numbers should give us a more quantitative appreciation of the value of this recently discovered astronomical windfall. Transporting a pound of mass from the Moon to L5 requires about a million foot pounds of energy, which is equivalent to 0.38 kilowatt hours.

Assuming that this will be accomplished by a Moon based mass driver, and generously allowing only a factor of three loss — for inefficiencies due to a) imperfect lunar / L5 attitude, b) imperfect solar collector / Sun attitude, and c) imperfect solar radiation to electricity conversion and electricity to kinetic energy conversion

— a solar collector with an area of a square kilometer can send one million pounds per hour from the Moon to L5. (Let's temporarily ignore that once this mass arrives in L5's vicinity, additional energy will be required to establish the stable orbit promised by Lagrange.) At this rate, the 4 million tons required to build an *Island One* can be transported in approximately a year. Transporting from the Moon a mass equivalent to 10% of the available Earth crossing asteroids for *Island One* designs would require 10,000 years. Transporting the mass for 10% of the small cities described above would require another 20,000 years. Clearly, depending only on the Moon / mass driver scheme will fall far short of accomplishing two of O'Neill's fondest hopes: achieving a growth rate of wealth that is greater than humanity's population growth, and anything even approaching an exponential growth of colonization. The far more robust strategy is to take advantage of the mass that already exists in high orbit and which, as John Lewis amply explains, provides substantially more diversity of material than that available on the lunar surface.

As diverse as the asteroidal materials may be, it's not certain that they will provide enough water to support all of these new biospheres. To avoid exporting water from the Earth, or from the surface of the Moon if the preliminary indications of ice bear fruit, we would look to the comets.

Unfortunately, the dynamics of comets are far less convenient for our purposes than those of asteroids. Their kinetic energies are an order of magnitude greater and their orbits often extend beyond the farthest planet. However, the orbits of many comets have been perturbed over the billions of years of their history into lower energy, asteroid-like orbits. It is estimated that about half of the near Earth orbit asteroids are extinct comets. Even one of the 1 km fragments of Comet SL9 which slammed into Jupiter in 1994 could furnish enough water for thousands of space habitats. For this type of resource, the exploitation will probably be shared by a community of habitats. The water content of 1 m to 10 m cometary fragments with low-to-medium energy orbits will probably be high enough that they would be far more economical to exploit than exporting water from the Earth or Moon.

Which brings us back to a point mentioned earlier: what will all these good folks in space *do?* What motivations — especially commercial motivations — will spur the beginnings of these migrations? O'Neill was optimistic regarding the value of beaming power from space to Earth, but rather pessimistic regarding the value of transporting material resources from space to Earth. As was mentioned many times already, he was unaware of the magnitude of these resources. Many detailed studies have been performed on the high value in today's marketplace of asteroidal materials, and John Lewis' work has been augmented by Mark Sonter at the University of Wollongong, by James Benson of the Space Development Corporation, by Jeffrey Kargel of the U.S.G.S, and by many others. So as the asteroidal space colonists are busy building their new civilization(s), they will pay their initial costs by exporting both power *and* materials to Earth. This part of the analogy with the colonization of the western hemisphere will be accurate, in my opinion. And the analogy can be extended: although the asteroid material diversity will be far higher than that of the lunar or Martian surfaces, it can be expected — as on Earth — that at any one location it will not be high enough and there will be vigorous commerce between most of the habitats. At some point it can be anticipated that the inter-habitat commerce will exceed commerce with the Earth and perhaps it will then be claimed that humanity will have achieved a permanent, stable civilization in space.

Before we rush out to buy tickets and sell our homes to buy space property, we must consider the many steps required to achieve these goals — many of which were addressed in the *High Frontier,* and in the SSI Space Manufacturing Conferences held at Princeton over the past 20 years, and by SSI sponsored research programs. One key nontrivial task that must logically precede all the others is to acquire and determine the orbits of this asteroidal population. Unfortunately, most modern astronomers concentrate on planetary, interstellar and intergalactic phenomena, and only a tiny handful worldwide are interested in the planetesimals of near Earth space — according to NASA's David Morrison, less than the staff of a single McDonald's restaurant. Presently, only about one-third of the population of the large asteroids mentioned above and far less of the small asteroids and comets have been acquired and catalogued. It would take many decades to increase our knowledge to 90% if this activity remained at the present pace. Although these objects pass relatively close to the Earth, astronomically speaking, their small size and low albedo make them extremely difficult to detect, especially against the galactic starry background.

Fortunately, there is a growing movement which calls itself *Planetary Defense* which is also strongly advocating the acceleration of acquiring and cataloguing Earth orbit crossing objects. Based on admirable interdisciplinary research advances among the world's astronomers, geologists, paleontologists and atmospheric physicists, it is now generally believed that the cause of the extinction of the dinosaurs and two thirds of all species then alive 65 million years ago was the Earth impact of an asteroid or comet. Moreover, there have been other impacts — some with even greater effect — and the hazard, however low the annual probability, continues today. Even an object an order of magnitude smaller than the dinosaur killer could cause tens of millions of casualties and trillions of dollars in damage, threatening our fragile infrastructures. Presently there are no known large objects heading Earth's way, but we know so little about the specific orbits that the first prudent step — as in all good risk management programs — is to define the potential hazard quantitatively. SSI endorsed a position paper, along with the American Institute of Aeronautics and Astronautics (AIAA), the Institute of Electrical and Electronic Engineers (IEEE) and the International Council on Systems Engineering (INCOSE), that was sent to Congress urging that a program to accelerate the acquisition and accurate orbit determination of near-Earth objects (NEO's) by at least a factor of ten. The response was somewhat slow but essentially positive: NASA substantially increased its budget and established a central management focal point for this purpose. The US Air Force is committing some of its military assets for a cooperative program with JPL on Maui, and the use of advanced electronic focal plane arrays integrated with computers provides a far greater sensitivity and image extraction effectiveness than the previous approach based on photographic film. Anticipating the results of this survey after a decade or two of hard work, there will be far less than one chance in a thousand that a 1 km asteroid will be found on an intercept trajectory with Earth, but most of the asteroidal population mentioned above will be acquired and their orbits catalogued.

As we struggle to determine the proper pace and sequence of activity in solving our complex puzzle, it's appropriate to reflect on unfolding events and on technologies that are advancing at diverse rates, and reconsider what is easy and what is hard. In my opinion, the hardest activity is sending people to live in space safely, comfortably and productively. Especially hard is to send families out in little spacecraft to search the unknown reaches of space for little homesteads — the way that families in covered wagons pushed the American frontier westward in the 19th century. It would be much easier to send mass, equipment and telerobots ahead of us to pave the way. And even on manned space operations the increased use of telerobots is surely justified over the previous man / machine tasking allocations. Easier yet is the transmission of energy via solar collectors and power beaming. This is so much more efficient than lugging around expendable chemical fuel, which itself presents much more of a propulsive burden than the payload it's used to deliver. Easiest of all is the transmission of information. Half of SSI's most difficult research program — the design of a self-replicating system — can be solved by transmitting the "genetic code" to a new generation rather than have the "parents" pass the code to its "progeny." Edward Teller is famed for his respect of high energy devices — however, he told the 1995 Planetary Defense Workshop that "Information is sometimes far more powerful than high energy." He was speaking of the "brilliant mountains" concept for intercepting hazardous objects — if we know the orbits of the planetesimals with sufficient accuracy, it will be far more effective to steer a small object into the path of a larger threatening object than to attempt to blow it up with a thermonuclear device. A more practical application is described by Freeman Dyson in *Imagined Worlds,* which involves a dramatic reduction in the energy required to deflect a threatening asteroid by concentrating on accurate information about its orbit, thereby increasing the warning time before impact. These concepts of employing information to massively reduce energy requirements can be extended to the high frontier.

The Sun radiates a billion times more energy than what is intercepted by Earth, allowing us to make the claim that space offers a virtually unlimited supply. However, as steady and dependable as this power is, it's too diffuse for many applications. As mentioned earlier, we presently find it far more practical to burn fossil fuels — which were formed from solar energy — at a rate a million times more rapidly than they were deposited. Quite often the intensity of power usage differs from the power available by orders of magnitude. We need to think through carefully the system aspects of power gathering, usage and storage. Also, if we think carefully about Teller's advice, there's an energy source available in the solar system that is far richer than solar power — and no, it's *not* nuclear power. It's the kinetic energy of the trillion or so planetesimals in the inner solar system. Even an object a fraction of a meter in size has enormous kinetic energy as it whizzes by at many km/sec. If a future interplanetary / inter-habitat navigator had a thorough and accurate map (a dynamic map in his computer of course), a direct mass driver (linear motor), a reverse mass driver (linear generator) and long tethers, all sorts of energy saving stratagems would be possible. These operations can include:

a) navigating by conventional chemical propulsion to the vicinity of a planetesimal with a known kinematic state and performing a momentum exchange maneuver with a tether and reverse mass driver, so that not only will the ship achieve a much more favorable velocity vector, but an additional surge of energy can be transferred into the ship's storage due to the tether's driving the reverse mass driver,

b) the rendezvous of two ships with different destinations in the above manner where each ship gets a momentum boost in the desired direction while also adding a surge of energy to storage,

c) the propelling of mass from one point and trajectory to another point using O'Neill's original scheme, and the capturing of that mass at the destination point, putting it into a new desired orbit by the use of a reverse mass driver, again gaining energy in the process most of the time, and

d) the establishment of a power station which has no particular destination but visits passing planetesimals for the sole purpose of gathering surges of energy, and eventually delivers the stored power to a habitat or major construction site where high power is required for short durations.

In order for these schemes to have any chance of working, a high degree of accurate and timely information will be required. As comfortable as life inside the habitats promises to be, the life of these "energy jockeys" will be dynamic, fast, exciting and even acrobatic.

The degree to which space habitats can become independent colonies is still an open question. Although it's not absolutely necessary for the space habitats to operate as stable closed systems — there will be substantial commerce between them to assure sufficient supplies of any necessity — it would still be comforting to know that any given habitat could survive an emergency period of a year or more with limited or no outside contact. Unfortunately, our hopes that this would be an easily achievable task have not been supported by the research of the last decade. The high visibility but poorly designed Biosphere experiments were disappointing, but a body of other research, such as William Jewell's work on bioremediation, promises that a relatively high degree of independence will eventually be practical. Considering these uncertainties, the prudent path to the high frontier should initially stress *open* rather than closed systems, with a maximum practical interdependence of the entire society of space colonists, rather than the solitary, isolated outpost model. With the benefit of additional research and experience, independence will become increasingly commonplace, if desired.

In summary, many aspects of Gerry O'Neill's vision remain valid today, but there is still much to do. These elements of O'Neill's high frontier are as valid today as when he defined them over twenty years ago:

○ The exploitation and colonization of space is essential to humanity's material future as well as its sense of purpose and freedom.

○ Space resources should be employed rather than expensively lifting resources from Earth, and the highest initial priority should be given to space power beamed down for use on Earth.

○ Human colonization sites should emphasize rotating space habitats rather than the surfaces of the planets and moons of the solar system.

○ Efficient transducers of sunlight to kinetic energy should be developed to transport mass and to provide propulsion within the inner solar system.

Despite the accuracy of the vision of the high frontier, its pace of development has been disappointingly slow. There is much to be done, beginning with a critical examination of those tempting illusions that we share with most of the idealistic space enthusiasts: There's the illusion that US space policy and funding can be extrapolated from the early years of the Sputnik to Apollo era, and that the cost of lifting mass to orbit will be substantially reduced. There's the illusion that manned space missions are analogous to the exploration and colonization of the western hemisphere. There's the illusion of rationality — that good ideas will be implemented merely because they're good, and that it's not necessary to sell them to the decision makers. And there's the illusion of a catalytic ignition followed by exponential growth, where humanity and its machines self-replicat into the universe. Finally, there's the illusion that science fiction is an accurate predictor of the pace of technology and of the future of the human race.

There *have* been favorable trends and emerging technologies. We must recognize them and seize the opportunities:

○ The cold war is over. This diminishes the "space race" priority, but more resources are now available to improve the lot of humanity. Our challenge is to sell the high frontier to the public and to the governments as an excellent long term investment in the well being of all Earth's citizens. SSI and its sister organizations SFF and ProSpace, have been increasingly effective in getting this message to Congress and to the general public.

○ The concept of space solar power, after an unfortunate setback in the 1980's, is once again receiving substantial funding. The National Science Foundation has sponsored a special conference to evaluate promising SSP technologies and systems concepts.

○ Cheap Access to Space is enjoying increasing attention from both governments and private entrepreneurs. A factor of ten cost reduction is reasonable employing conventional rocketry and another factor of ten should be attainable employing advanced techniques such as mass drivers and laser beam rider propulsion.

○ Advances in microelectronic and micromechanical systems over the past decades permit the far more efficient utilization of payloads and telerobotics for a great variety of missions and new man / machine tasking strategies.

○ Our knowledge of the resources of near-Earth space — especially Earth orbit crossing asteroids (ECA's) — has vastly improved over the past ten years. Not only do the ECA's promise a far greater variety of crucial material than the lunar surface, many of them are more reachable from an energy expenditure standpoint. The logistics of building habitats now favors construction at the site of the asteroids, rather that transporting material from the Moon to L5.

○ The mass driver (linear motor) is an inspirational concept to free us from the traditional restrictions of rocketry, but it should be extended to include the mass catcher (linear generator) in order to complete the problem of mass transport, propulsion and momentum transfer within the inner solar system.

○ Sending people into space is our most difficult task. The use of telerobots, when practical, is far easier. Easier yet is the transmission of energy. Easiest of all is communication, the transmission of information. Our technology plans and investment strategies should be mindful of this ladder of difficulty. In the past they have not been.

So we see that the puzzle is truly complex and multidimensional, and there's more to the solution than science and technology, just as Gerry O'Neill predicted. Although the pace over the past two decades was far slower than he had hoped, there is far more good news than bad. What we must avoid most of all is the slippery slope of allowing our leaders to assume that the exploitation and colonization of space is so far in the distant future that it can be safely postponed until we take care of today's needy populations. Space is the only viable long term solution for those needy populations, and if we can't afford to invest in it today then it will become even less affordable with the relentless growth of these populations. The high frontier vision is still valid and new technologies, such as cheap microelectronics; new engineering advances, such as cheap access to space; and new discoveries, such as the planetesimals; can be constructed into a fresh set of strategies which find new balances between transporting people or equipment, and transmitting energy or information. We "merely" need to be faithful to our ideals, not lose hope, work harder and be more articulate as we communicate our vision to a broader constituency.

When I was a young boy during the depression, my family was living on my father's disability pension in upstate New York. We were clearly at the low end of the lowest income bracket, not even able to afford a newspaper, much less a telephone or car. To survive, we had to religiously conserve every resource. We produced virtually no garbage and I was painfully embarrassed that I had nothing to bring to the annual school paper drives. This was our credo:

> Use it up,
> Wear it out;
> Make it do,
> Do without.

If humanity continues on its present path, and hopefully avoids food riots and major wars for resources, it will still eventually be impoverished compared to our present state. Even the children of the *richest* men on Earth will recite this credo every morning. Both our minds and our bodies will be starved.

Mankind must become a spacefaring race. We will go for the science; we will go for the thrill of adventure, of being explorers and pioneers; we will go for the freedom and challenges of new frontiers; we will go for the riches; we will go for the fulfillment of our destiny and the uplifting of our spirit. But most of all, we will go because we have to.

# References

**Chapter 1**

1. Lucian of Samosata, *A True History* and *Icaromenippus*, circa A.D. 160, English translations (respectively), New York: Murray, Scribner & Welford, 1880, and Oxford: Clarendon Press, 1905.
2. E. E. Hale, "The Brick Moon," *Atlantic Monthly*, vol. XXIV, October, November and December, 1869.
3. J. Verne, *Off on a Comet*, Paris, 1878.
4. K. K. Lasswitz, *On Two Planets*, Leipzig, 1897.
5. K. E. Tsiolkowsky, *Dreams of Earth and Heaven*, Moscow, 1895.
6. K. E. Tsiolkowsky, *The Rocket into Cosmic Space*, Moscow, Naootchnoye Obozreniye, 1903.
7. R. H. Goddard, "The Ultimate Migration," manuscript dated Jan. 14, 1918, The Goddard Biblio Log, Friends of the Goddard Library, Nov. 11, 1972.
8. R. H. Goddard, "2. Importance of Production of Hydrogen and Oxygen on the Moon and Planets," manuscript notes, March 1920.
9. H. Oberth, *The Rocket into Interplanetary Space*, Munich, 1923.
10. G. von Pirquet, articles, *Die Rakete*, vol. II, 1928.
11. H. Noordung (Potocnik), *The Problems of Space Flight*, Berlin: Schmidt and Co., 1929.
12. J. D. Bernal, *The World, the Flesh and the Devil*, London: Methuen & Co., Ltd., 1929.
13. O. Stapledon, *Starmaker*, London: K. Paul, Trench, Trubner & Co., 1929.
14. H. T. Rich, "The Flying City," *Astounding Stories*, August 1930.
15. F. Zwicky, "Morphological Astronomy," The Halley Lecture for 1948, delivered at Oxford, May 2, 1948, *The Observatory*, vol. 68, August 1948, pp. 142-3.
16. H. E. Ross, "Orbital Bases," *J. British Interplanetary Society*, vol. 8, no. 1, 1949.
17. A. C. Clarke, "Electromagnetic Launching as a Major Contributor to Space Flight," *J. British Interplanetary Society*, vol. 9, 1950, p. 261.
18. W. von Braun, "Crossing the Last Frontier," *Collier's*, March 22, 1952.
19. L. R. Shepherd, "Interstellar Flight," *J. British Interplanetary Society*, July 1952.
20. A. C. Clarke, *Islands in the Sky*, Philadelphia: John C. Winston, 1952.
21. I. M. Levitt and D. M. Cole, "Exploring the Secrets of Space," Englewood Cliffs, N. J.: Prentice Hall, Inc., 1963, pp. 277, 278.
22. F. J. Dyson, "Search for Artificial Stellar Sources of Infrared Radiation," *Science*, vol. 131, June 1, 1960, p. 1667.
23. V. P. Petrov, *Artificial Satellites of the Earth*, translated by B. S. Sharma and R. S. Varma, Ministry of Defense, Gov't of India, New Delhi: Hindustan Publishing, 1960, p. 217.
24. K. P. Osminin, "Questions of Economics and International Cooperation in Space Operations," XXVth International Astronautical Congress, Amsterdam, The Netherlands, Sept. 30-Oct. 5, 1974.
25. K. A. Ehricke, "Space Stations — Tools of New Growth in an Open World," XXVth International Astronautical Congress, Amsterdam, The Netherlands, Sept. 30-Oct. 5, 1974.
26. A. Berry, *The Next 10,000 Years*, New York: Saturday Review Press/E. P. Dutton & Co., 1974.
27. G. Harry Stine, The Third Industrial Revolution, New York: G. P. Putnam's Sons, 1975.

(For references 1-25, cf. R. Salkeld, "Space Colonization Now," *Aeronautics and Astronautics*, September 1975, p. 30, as reviewed by F. C. Durant III.)

**Chapter 2**

1. Carleton S. Coon, *The Story of Man*, New York: Alfred A. Knopf, 1954.
2. Sebastian von Hoerner, "Population Explosion and Interstellar Expansion," in *Einheit und Vielheit*, Göttingen: Van den Houck & Ruprecht, 1973.
3. J. C. Fisher, *Energy Crises in Perspective*, New York: John Wiley & Sons, 1973.

4. E. F. Schumacher, *An Economics of Permanence*, Institute for the Study of Non-Violence, Box 1001, Palo Alto, California 94302.
5. Population Studies #53, U.N. Department of Economic and Social Affairs, United Nations, New York, 1973.
6. Von Hoerner, *op. cit.*
7. *Ibid.*
8. P. A. Taylor, "World Population Conference 1974"; interview with Ansley J. Coale: *Princeton Alumni Weekly*, Oct. 22, 1974, p. 8.
9. David R. Safrany, "Nitrogen Fixation," *Scientific American*, October 1974, vol. 231, #4, pp. 64-81.
10. Fisher, *op. cit.*
11. Associated Universities, Inc. AET-8, April 1972.
12. Fisher, *op. cit.*
13. Jean-Jacques Faust, *L'Expresse*, Nov. 18-24, 1974.
14. Safrany, *op. cit.*
15. Associated Universities, Inc., *op. cit.*
16. J. McPhee, "The Curve of Binding Energy," *New Yorker*, Dec. 17, 1978.
17. Von Hoerner, *op. cit.*
18. Schumacher, *op. cit.*
19. Robert Heilbroner, *An Inquiry Into the Human Prospect*, New York: W. W. Norton, 1974. Page references: 134, 17, 88, 44, 43, 93, 110, 108, 26, 141, 140, 27, 136.
20. J. W. Forrester, *World Dynamics*, Cambridge, Mass.: Wright Allen Press, 1971.

## Chapter 3
1. *National Geographic*, January 1975.
2. Gerald Feinberg, *The Prometheus Project*, Garden City, New York: Doubleday, 1969.
3. David Hafemeister, "Science and Society Test for Scientists: The Energy Crisis," *American Journal of Physics*, August 1974.

## Chapter 4
1. David R. Safrany, *op. cit.*
2. J. C. Fisher, *op. cit.*
3. G. Harry Stine, *op. cit.*
4. Lewis Beman, "Betting $20 Billion in the Tanker Game," *Fortune*, August 1974.
5. T. B. McCord and M. J. Gaffey, "Asteroids: Surface Compositions from Reflection Spectroscopy," *Science*, October 1974.
6. E. K. Gibson, C. B. Moore, and C. F. Lewis, "Total Nitrogen and Carbon Abundances in Carbonaceous Chondrites," *Geochimica et Cosmochimica Acta*, 35:599, June 1971, #6, pp. 599-604.
7. K. Tsiolkowsky, *Beyond the Planet Earth*, trans. by Kenneth Syers, New York: Pergamon Press, 1960
8. K. Tsiolkowsky, "Selected Works," Moscow: Mir Publishers, 1968.

## Chapter 5
1. G. K. O'Neill, "Colonization at Lagrangia," *Nature*, August 23, 1974.
2. G. K. O'Neill, "The Colonization of Space," *Physics Today*, September 1974.
3. "Multiple Cropping — Hope for Hungry Asia," *Reader's Digest*, October 1972, p. 217.
4. Richard Bradfield, private communication.
5. F. M. Lappe, *Diet for a Small Planet*, New York: Ballantine Books, 1971.
6. "Multiple Cropping . . .," *op. cit.*

## Chapter 6
1. G. K. O'Neill, "The Colonization . . .," *loc. cit.*
2. Henry H. Kolm and Richard D. Thornton, "Electromagnetic Flight," *Scientific American*, October 1973.
3. Arthur C. Clarke, "Report on Planet Three," Signet #5409, New York: New American Library, 1972.
4. Richard Bach, Dedication to *Jonathan Livingston Seagull*, New York: Macmillan Co., 1970.

## Chapter 7

1. "Meteoroid Environment Model — 1969 (Near Earth to Lunar Surface)," NASA SP-8013, 1969.
2. G. Latham, J. Dorman, et al., "Moonquakes, Meteorites and the State of the Lunar Interior," and "Lunar Seismology," in *Abstracts of the Fourth Lunar Science Conference*, 1973; Lunar Science Institute, 3303 NASA Road 1, Houston, Texas 77058.
3. R. E. McCrosky, "Distributions of Large Meteoric Bodies," Smithsonian Astrophysical Observatory Special Report #280, 1968.
4. Morgan and Turner, eds., *Natural Environment Radiation Exposure*, New York: John Wiley & Sons, 1967.
5. G. M. Comstock, R. L. Fleischer, et al., "Cosmic Ray Tracks in Plastics: The Apollo Helmet Dosimetry Experiment," *Science*, April 9, 1971
6. *Wissenschaften* 60, 233, 1973.
7. John McPhee, *The Curve of Binding Energy*, New York: Farrar, Straus & Giroux, 1974.

## Chapter 8

1. Edison Electrical Institute, *Statistical Yearbook of the Electric Utility Industry for 1973*, New York: Edison Electrical Institute, 1973.
2. G. D. Friedlander, Institute of Electrical and Electronic Engineers *Spectrum 12*, May 1975.
3. W. R. Cherry, *Aeronautics and Astronautics*, August 1973.
4. Report of the 1975 NASA-Ames/Stanford University Summer Study on Space Colonization.
5. H. Davis, *Proceedings*, 1975 Princeton University Conference on Space Manufacturing Facilities, Paper I-6, New York, American Institute of Aeronautics and Astronautics (AIAA).
6. A. O. Tischler, *ibid.*, Paper 1-5.
7. G. K. O'Neill, "The Colonization . . .," *loc. cit.*
8. A. C. Clarke, *Journal of the British Interplanetary Society*, vol. 9, 1950.
9. H. H. Kolm and R. D. Thornton, "Electromagnetic Flight," *loc. cit.*
10. Kevin Fine, Eric Drexler, Bill Snow, Jonah Garbus (M.I.T.) Jon Newman (Amherst).
11. B. Mason and W. G. Melson, *The Lunar Rocks*, New York: Wiley-Interscience, 1970.
12. G. K. O'Neill, "The Low (Profile) Road to Space Manufacturing," *Astronautics and Aeronautics*, September 1977.

## Chapter 9

1. D. Hayes, *Science*, June 27, 1975.
2. W. R. Cherry, "Harnessing Solar Energy: The Potential," *Aeronautics and Astronautics*, August 1973.
3. P. E. Glaser, Space Shuttle Payloads, Hearing, Committee on Aeronautical and Space Sciences, U. S. Senate, Oct. 31, 1973, Part 2.
4. W. C. Brown, Proc. IEEE, January 1974.
5. News release, Office of Public Information, Jet Propulsion Laboratory, California Institute of Technology, Pasadena, May 1, 1975.
6. G. L. Woodcock and D. L. Gregory, American Institute of Aeronautics and Astronautics paper 75-640, presented at the American Institute of Aeronautics and Astronautics/American Astronomical Society Conference on Solar Energy for Earth, April 24, 1975.
7. K. Bammert and G. Deuster, paper presented at the American Society of Mechanical Engineers Gas Turbine Conference, Zurich, April 1974.
8. R. E. Austin and R. Brantley, presentation at NASA Headquarters, Washington, D.C., April 17, 1975 (unpublished).
9. R. E. Austin, NASA-Marshall Spaceflight Center, private communication.
10. G. K. O'Neill, "Space Colonies and Energy Supply to the Earth," *Science*, Dec. 5, 1975.
11. G. K. O'Neill, Testimony before the Subcommittee on Space Science and Applications of the Committee on Science and Technology, U. S. House of Representatives, July 23, 1975. Superintendent of Documents, U. S. Government Printing Office, Washington, D. C. 20402.
12. G. K. O'Neill, Power Satellite Construction from Lunar Surface Materials, Testimony before the Subcommittee on Aerospace Technology and National Needs of the Committee on Aeronautical and Space Sciences, U. S. Senate, January 19, 1976; U. S. Government Printing Office, Washington, D. C. 20402.
13. Exxon Corporation, advertisement in *Smithsonian* magazine, April 1975.
14. G. K. O'Neill, "A High Resolution Orbiting Telescope," *Science*, May 1968.

15. R. N. Bracewell, "The Galactic Club," Stanford Alumni Association, Stanford, California, 1974, and *Nature*, vol. 186, 1960.
16. G. Cocconi and P. Morrison, "Searching for Interstellar Communications," *Nature*, vol. 184, 1959.
17. F. D. Drake, "Project Ozma," *Physics Today*, vol. 14, p. 140, 1961.
18. Ian Ridpath, *Worlds Beyond*, New York: Harper & Row, 1975.
19. R. N. Bracewell, *op. cit.*
20. Bernard M. Oliver, ed, "Project Cyclops," NASA Report #CR114445, 1973.
21. Carl Sagan, *The Cosmic Connection*, New York: Doubleday-Anchor Press, 1973.

## Chapter 10
1. T. Taylor, "Propulsion of Space Vehicles," in R. Marshak's *Perspectives in Modern Physics*, New York: Wiley-Interscience, 1966.
2. R. Bradfield, "Multiple Cropping . . .," *loc. cit.*
3. F. M. Lappe, *loc. cit.*

## Chapter 11
1. H. S. F. Cooper, Jr., *A House in Space*, New York: Holt, Rinehart & Winston, 1976.
2. T. B. McCord and M. J. Gaffey, *loc. cit.*
3. C. R. Chapman, D. Morrison, and B. Zellner, "Surface Properties of Asteroids: A Synthesis of Polarimetry, Radiometry and Spectrophotometry," *Icarus*, vol. 25, 1975.
4. *Ibid*
5. For turbogenerator combinations in the 1300 MW size range, the purchaser (TVA) in 1975 paid the builder (Brown-Boveri Corp.) $56/KW, (*Wall Street Journal*, Jan. 23, 1975). The corresponding cost per 500 KW is $28,000. In the text, the figure is doubled to include nonturbogenerator components of a complete power plant.
6. R. Heilbroner, *loc. cit.*, p. 140.

## Chapter 12
1. Klaus P. Heiss, "Our R+D Economics of the Space Shuttle," *Aeronautics and Astronautics*, October 1971.
2. G. K. O'Neill, *Nature, loc. cit.*
3. Columbus lived from 1451 to 1506; Francis Drake, from 1545 to 1595; Michelangelo spanned the years 1475-1564, and Shakespeare 1564-1616

## Chapter 15
1. Fuller, B. and Kurimaya, C., "Critical Path", 1981.
2. Glaser, P.E., "Power from the Sun: Its Future," Science, November 22, 1968, vol. 162, pp. 857-861.
3. Glaser, P.E., Davidson, F.P., and Csigi K.I., "Solar Power Satellites — A Space Energy System for Earth," John Wiley & Sons Ltd., U.K., ISBN 0471 96817X, 1997.
4. Duncan, R.C. and Youngquist, W., "Encircling the Peak of World Oil Production," Natural Resources Research, Vol. 8, No. 3, 1999.
5. Glaser, P.E., "Wireless Power Transmission From Space To Offshore Receiving Antennas To Meet Ocean Cities Energy Needs," Presented at Cite Marines '95, Fondation 2100', SEE, Monaco, November 20-23, 1995.
6. Glaser, P.E., "Power Without Pollution", 5th International Microwave Power Institute, Scheveningen, Netherlands, October 8, 1970.
7. NASA and U.S. Department of Energy, "Satellite Power System Concept Development and Evaluation Program, System Definition Technical Assessment Report," DOE/ER/10035-03, December, 1980.
8. NASA, HQ., "An Executive Summary of Recent Space Solar Power Studies and Findings, Concepts, System Analysis, Space Mission Applications, and Technology Road Mapping," December, 1998.
9. Canadian Department of Communications, "Stationary High Altitude Relay Platform, (SHARP" Ottawa, Canada, 1987.
10. Kobe University, "Microwave Powered Aircraft (MILAX)", August 1992.
11. Glaser, P. E., "The Power Relay Satellite", Plenary Lecture, 44th Congress of the International Astronautical Federation, October 16-22, 1993, Graz, Austria.

12. *Space Policy*, Special Issue on SSPS, published by Elsevier Science Ltd., The Boulevard, Langford Lane, Kidlington, Oxford, UK, OX5 1GB, England, May 2000.
13. O'Neill, G.K., "2081 — A Hopeful View of the Human Future", Simon & Schuster, New York, copyright 1981 by Gerard K. O'Neill, ISBN 0-671-24257-1, page 268.

**Chapter 18**
- Gerard K. O'Neill, *The High Frontier, 2nd Edition, SSI* Press, Princeton, NJ, 1989
- Gerard K. O'Neill, *Space Resources and Space Settlements,* NASA 1977 Summer Study, NASA SP-428
- Robert A. Freitas, Jr., *Advanced Automation for Space Missions,* NASA 1980 Summer Study, NASA CP-2255
- Freeman J. Dyson, *Imagined Worlds, Harper* Collins, 1996
- John S. Lewis, *Mining the Sky, Untold Riches from the Asteroids, Comets and Planets,* Addison Wesley, New York, 1996
- Alberto Behar, *On the Design of Sub-Kilogram Intelligent Telerobots for Asteroids; PhD Dissertation, University* of Southern California, May 1998
- Mark Sonter, *The Technical and Economic Feasibility of Mining the Near-Earth Asteroids, MS Thesis, University* of Wollongong, Australia, 1997
- Edward O. Wilson, *Consilience; The Unity of Knowledge, Alfred* A. Knopf, New York, 1998
- Zubrin, Robert, *The Case for Mars, Simon* and Schuster, New York, 1996
- George Friedman, Risk *Management Applied to Planetary Defense,* IEEE Transactions on Aerospace and Electronics Systems, V. 33, No.2; April 1997
- George Friedman, *Responding to the Threat of a Near-Earth Object Impact, AIAA* Position Paper, Washington, DC, Aug 1995
- Tom Gehrels, ed, *Hazards due to Comets and Asteroids,* U. of Arizona Press, Tucson 1995
- David Morrison, Ed, *The Spaceguard Survey, Report of the International NEO Detection Workshop, Office* of Space Science and Applications, NASA, Washington, DC, 1992

# Index

Space Studies Institute
P.O. Box 82
Princeton, NJ 08542

| Phone: | (609)-921-0377 |
| FAX: | (609)-921-0389 |
| E-mail: | ssi@ssi.org |
| Web: | http://www.sii.org |

Space Frontier Foundation
Mailing Address:
    8391 Beverly Blvd. #493
    Los Angeles, CA 90048

Street Address:
    11350 Ventura Blvd. Suite 100
    Studio City, CA 91604

| Phone: | (818) 985-7781 |
| | (800) 78-SPACE |
| Fax: | (818) 985-8767 |
| Email: | information@space-frontier.org |
| Web: | http://www.space-frontier.org |